iWar

www.**penguin**.co.uk

ALSO BY TIM HIGGINS

Power Play: Tesla, Elon Musk, and the Bet of the Century

iWar

Fortnite, Musk, Spotify, and the Siege of Apple

Tim Higgins

torva

TRANSWORLD PUBLISHERS

UK | USA | Canada | Ireland | Australia
India | New Zealand | South Africa

Transworld is part of the Penguin Random House group of companies
whose addresses can be found at global.penguinrandomhouse.com.

Penguin Random House UK, One Embassy Gardens,
8 Viaduct Gardens, London SW11 7BW

penguin.co.uk

Penguin
Random House
UK

First published in Great Britain in 2025 by Torva
an imprint of Transworld Publishers

001

Designed by Michele Cameron
Printed and bound in India by Manipal Technologies Limited

The authorized representative in the EEA is Penguin Random House Ireland,
Morrison Chambers, 32 Nassau Street, Dublin D02 YH68.

A CIP catalogue record for this book is available from the British Library.

ISBNs:
9781911709640 hb
9781911709657 tpb

For Karin

CONTENTS

AUTHOR'S NOTE

This book was made possible by people involved with these companies and government agencies that talked to me—often on the condition that those conversations not be attributed to them directly. Some of the people mentioned by name agreed to be interviewed, some declined. To protect my sources, who risked professional reprisal, I don't generally disclose who I interviewed. This book also relies heavily on reams of internal documents from these companies made public through litigation and hours and hours of legal testimony. In a few instances, direct quotes were recreated from my sources' memories to the best of their recollection. These quotes were heard firsthand by at least one source. Most of the time, though, direct quotes came from documents, such as emails, reviewed by me, or audio and video recordings. I conducted reporting in Austin, Cupertino, Brussels, New York City, San Francisco, and Washington, D.C. The main characters in this book were offered a chance to comment on details in this narrative.

PROLOGUE

2021

Tim Cook sat between two worlds. And his power over both was being tested. Seated in the back of an Oakland courtroom waiting to face a judge, the sixty-year-old man from Alabama who rose to chief executive of Apple did what we all tend to do in such tedious moments: looked at his iPhone. Unlike the rest of us, in that digital world Cook was the king, ruling over an iPhone empire with a billion users, all connected by the software that his company had developed, shipped, policed, and taxed.

The real world wasn't the same since the iPhone was revealed fourteen years earlier in 2007. Communities were reimagined. Commerce evolved. Giants rose on the success of a new computing paradigm. And now, in the spring of 2021, Cook was in the real world of a federal court, having to answer for the power he held in his digital realm. He came prepared to defend against those who wanted to strip away the rules and order, who wanted to tear down the Walled Garden that, he and his team insisted, kept their world special and that users wanted.

As Cook took the stand, wearing a dark gray suit, white shirt, and gray tie, a clear face shield protected him from germs, part of the unusual safeguards of the pandemic era, precautions that included limiting who could be in Judge Yvonne Gonzalez Rogers's courtroom that Friday in May. Because of that limited access, she took the rare step of allowing audio of the proceedings to be broadcast to an army of lawyers, regulators, and rivals listening around the world. They wanted to hear Cook defend Apple's reign over the App Economy.

With his polite southern drawl, Cook could, at times, sound both awestruck and personally offended at any suggestion his company would—*could*—do wrong. The digital world that Apple created with its App Store was special. Apple was making the world better. "I think it's been an economic miracle," Cook told the judge from the witness stand.

That miracle had made Cook a billionaire and Apple the world's most valuable company. With annual sales approaching $400 billion, Apple's output exceeded most nations' GDPs. In 2021, it printed more than $3,000 in profit every second of every day. Yet this pinnacle had made Apple vulnerable, both to corporate rivals and to governments that increasingly questioned the power the company wielded over its digital empire—and them.

A rebellion against the empire had been brewing for a long time—almost from day one when it became clear how powerful the mobile computing world was going to be. But now things had escalated. Here, Cook was on the witness stand to defend Apple over claims that it was a monopolist. He was joined in the courtroom by a top deputy named Phil Schiller, a longtime Apple executive whose lifework amounted to the stewardship of the App Store and Apple's reputation.

Across the courtroom from Cook and Schiller sat the man who brought the lawsuit that was attracting the world's attention and had put Cook in the crosshairs: Tim Sweeney.

The iPhone was a gateway for his company, Epic Games and its wildly popular video game *Fortnite*, to wealthy consumers. One billion users depended on Apple's grace. And Cook's patience had run thin the previous summer, when Epic sprang a trap designed to draw his wrath and end up in this courtroom on this day to air out what Sweeney felt was Apple's dirty laundry. In the years since Apple had created its digital empire, with the advent of the iPhone, in order for other businesses to gain access to its marketplace, Apple had set up a drawbridge for all of the companies that wanted to make money through it. Unlike users of a computer that can download programs at will, iPhone users had to get third-party programs from one spot, Apple's App Store. To be offered in the App Store, Epic and other software makers agreed to certain rules around content and to use Apple's payment system, which conveniently collected as much as 30 percent of the revenue generated through each app. Cook and his team were essentially taking 30 cents of every dollar spent on digital goods in their empire, akin to a tax for breathing their air. He would argue it was only fair. That most apps were free and didn't have to pay for the opportunity to be part of the iPhone empire. Those who were making money selling digital goods and services, such as video games and streaming music, should pay their fair share. Apple, after all, was connecting those third-parties with its massive audience. Its rules were there to protect Apple's users from tricksters, hackers, and scammers. By its own calculations, Apple had stopped more than $1.5 billion in potentially fraudulent transactions in the previous year alone while rejecting hundreds of thousands of proposed apps for privacy violations or for being spam, copycats, or misleading. For Schiller, the attacks on Apple could feel personal. He had worked with the late Steve Jobs to build the App Store and was a true believer in the idea that they were offering a better experience for users—those customers of the iPhone and the small developers who could benefit from their economic miracle. He held tight to the lessons learned from his friend. "I know it sounds like bullshit . . . these philosophies—about small developers and about trying to do the right thing and not caring about money," Schiller told me. "To me, it's always honoring the things he taught us. I really want to hang on to as much of that as I can."

Those selling digital goods couldn't just go someplace else if they didn't like Apple's rules. Apple and its rival Google controlled the mobile world. Their rules were very similar, though Apple was often the lead dog, taking positions that Google followed. Some have long seen Apple's toll taking as a violation of U.S. monopoly laws that dated back to the era of robber barons. But few had the guts or the gold to challenge Apple. Sweeney would later estimate that his battle, which saw *Fortnite* booted from both Apple and Google app stores, cost his company more than $1 billion in both lawyer fees and lost sales. Facebook co-founder Mark Zuckerberg, of all people, had warned of

Apple's power, but had failed to diminish it. Daniel Ek, the founder of Spotify, had antagonized Apple for years as he sought to sell his streaming-music beyond Apple's reach, to moderate effect.

Months before Cook appeared in court, Apple's control of the App Economy took on a new political dimension after it booted a social media platform called Parler that had grown popular with conservatives who felt like Silicon Valley was trying to limit their speech. Apple, along with Google, ejected Parler over content moderation concerns following the January 6 Capitol insurrection. Soon, Elon Musk, the world's richest man, would face the limits of his own power when it came to Apple after acquiring Twitter amid worries liberals were hampering its role as a town square. According to Musk, Apple would quietly deliver a message that his social media platform could soon find itself withheld from the App Store—an existential threat for the little blue bird.

What was so jarring in 2021 was that the forces against Apple appeared to be closing in—a dramatic change from just years earlier when it was Apple's rivals, Google and Facebook, that were commonly seen as the Big Tech companies having gone too far. Now Apple, to its dismay, was being lumped in with something of a rogues' gallery.

The pressures mounting against Apple were not without historical precedent in American business history. With the rise of a new technology, power players have emerged that benefit from their scale and ecosystem's rules that seemingly locked in customers. That's been followed by decay as rivals and regulators tried to chip away at the power—sometimes successfully or other times not. But the fight tends to provide a distraction, allowing a new player armed with a newer technology to emerge. IBM spent much of the 1970s fighting the Justice Department's claims it maintained an illegal monopoly over computers only to see the Reagan administration drop that in the early 1980s. By then, Microsoft was on the rise, launching in 1985 its first version of the Windows operating system that would rise to such power as to attract its own antitrust claims a generation later. Apple's iPhone kingdom began to grow as Microsoft fought its monopoly battles in the U.S. and Europe.

By 2021, the list of Apple enemies was long. Together they formed a ragtag group of rebels fighting Apple for control of the internet—all of them engaged in an increasingly political fight. On this day, as Cook sat before a judge who would decide Apple's fate, it was Sweeney's turn to do damage, a gamble he could use the popularity of his game *Fortnite* to crack open Apple's Walled Garden.

The rebellion had been dragging on for years and would only grow—in part—because of the attention Sweeney brought. The Justice Department and Europe's top antitrust official were paying attention, as were Congress and the European Union. Apple's power had even inspired apparent aggressive maneuverings in China. It was a global story of conflicting interests, the entanglement of corporate powers, and geopolitics. It was a battle royal among some of the biggest names in tech. The motives of the rebels? Hardly pure. Billion-dollar fortunes were at stake. That, in part, helped Apple dismiss assaults upon it as sour grapes and vulgar greed. Apple has, it argues, simply built a better mousetrap. It has ushered into the world a product that rules the day, and has won consumer pocketbooks. Why shouldn't it benefit—in perpetuity?

Yet the tapestry of the rebellion's complaints has woven a new picture of a generational fight: Apple as a monopolistic power, an empire dictating to everyone commercial terms and, more crucially, cultural terms. Terms of expression. To them, in many ways, Apple set the limits of expression and thought. And as much as anyone might dislike those aligned against Apple, these people weren't exactly wrong.

PART I

NEW WORLDS

NEW WORLDS

1
AN UNREAL HERO

2006

The rare peek into Tim Sweeney's private life began at the front driveway of his mansion in the Raleigh, North Carolina, area. His orange Lamborghini roared toward the brick home. Pale, balding, and wearing an ill-fitting sweatshirt, Sweeney began by telling the MTV documentary, "I never really expected to be this successful when I went into game development."

Deeply private, Sweeney had been persuaded to appear on camera as part of a marketing push by Microsoft, which was betting his company's upcoming game—a gory military-meets-horror shooter—could help put its new gaming console, Xbox 360, on the map. Sweeney, who talks with a soft, almost tinny, high-pitched voice, proceeded to show the camera crew his bachelor pad's large dining room, which he confessed he never used, favoring to eat at fast-food restaurants such as Burger King. In his kitchen, he pointed to his collection of garnets, gathered during hikes; then he went out into the backyard, where the almost six-foot-tall, nearing forty-year-old man proceeded to climb a tree. "I have a really big house," he said. "I don't know why I have a big house. I really don't need it . . . but I figured, you know, I have the money, why not?"

It has been said that the video game industry is Hollywood for nerds. If so, in 2006, Sweeney was something of a Walt Disney in this world. His company, Epic Games, founded in 1991, debuted one of the hottest franchises in recent memory: *Gears of War*.

The MTV documentary was part of a marketing launch to build excitement for the game.

The 3D, over-the-shoulder-view shooter debuted on the Xbox 360 game system in late 2006. Within a few weeks, two million copies had been sold for the all-important holiday season. It was lauded with industry best-of awards and remained an Xbox 360 bestseller for almost a year. (A sequel to the game, *Gears of War 2*, would follow in 2008 and sell two million copies in the first weekend.)[1]

Attention followed. Glowing reviews. Stories in *Entertainment Weekly* and *The New Yorker* talked about his weird, private company based in suburban Raleigh—far from Silicon Valley. Sweeney wasn't an attention-getting personality like his star video game designer/ wunderkind Cliff "CliffyB" Bleszinski with bleached blond hair and knack for publicity. Sweeney was sincere, perhaps, at times, naive, and very determined. What you'd imagine if a teenager won the lottery—he had all the toys but was mostly just interested in his one passion. In fifteen years, Sweeney had managed to guide Epic through industry-upending technology shifts. The *Gears* franchise represented a culmination of a long bet that Epic could pivot from selling its software directly to customers through online bulletin boards to making its own sophisticated games for the big distributors like Microsoft. It had moved from a world of simple pixel graphics to immersive 3D wonders. If anything, Sweeney had shown a canny ability to see how technology was shifting and to position Epic to be ready.

Behind the *Gears* franchise was a software tool that Sweeney had led the creation of called Unreal Engine, which helped Bleszinski efficiently build his digital worlds. The first *Gears* game had a budget of around $12 million and a team of fewer than thirty people. Unreal allowed them— including nonprogrammers such as Bleszinski—to focus on the art and gameplay and not worry about the time-consuming process of building the pieces of their world, such as adding the landscapes or the physics of how characters might move. "It was as if Einstein had created a tool kit for the average person to understand space and time, capture it, and mold it into cool shapes that would appear on their computer screen," Bleszinski would later recall.[2]

Sweeney's passion seemed to lay in developing the tools of game making, in building a new world for others to play. He left the work of game development to artists such as Bleszinski, who had joined Sweeney as a teenager still in high school when Epic was operating out of his parents' basement. As the two young men worked on games, Bleszinski, would later write in his autobiography, he came to realize Sweeney "was more interested in making games than playing them."[3]

The success of the *Gears* franchise changed the trajectory of Epic and, in the process, made Sweeney, who was the largest shareholder in the private company, wildly rich—a far cry from his middle-class upbringing in suburban Maryland where the shy kid first learned to code.

At twenty years old, Sweeney had begun his video game company from his parents' home, initially calling it Potomac Computer Systems after his hometown in Maryland. The first game, *ZZT*, wasn't very good compared with the lifelike graphics of his games to follow years later. But it was a start. Players navigated a white smile-like character through mazes, interacted with clues, and shot enemies that looked like characters found on the keyboard. More important, perhaps, *ZZT* held the ingredients of what would ultimately make Sweeney successful in the years to come.

ZZT wasn't simply a dungeon-puzzle-style game; it also included a feature—an editor—that allowed users to make changes to the actual game, such as adding their own characters, or even make up a new game of their own within the digital world that Sweeney had created. That editor helped make the game popular among computer users during the days ahead of the modern internet.

At the time, Sweeney didn't realize it, but his desire to create tools that allowed others to easily create their own digital worlds would become a formula he would build upon again and again—one that would eventually put him at odds with Apple.

In 1991, however, Sweeney, still a student at the University of Maryland, wasn't thinking about changing the world. He was more interested in earning a buck, and until his experiment with *ZZT* he hadn't thought that possible. He was wrong, of course. He was on the path to great riches.

His interest in computers had begun at an early age, around eleven, after visiting one of his older brothers in San Diego. His brother Steve was more than fifteen years older and had moved across the country for a job developing hardware and software.

It was all pretty cool to the younger Sweeney, who idolized his brother and his new life that now included living in a small house near the beach, driving a nice car, and, perhaps best of all, spending his days working on computers. Beyond opening his littler brother's eyes to a world beyond suburban Washington, D.C., Steve gave Tim a life-changing gift: an Apple II Plus computer.

In the early 1980s, Apple was aiming to change the thinking around what a computer could be. Its co-founder Steve Jobs was selling the idea of a home computer, a personal gadget, not some big box used only by giant corporations. The Apple II was an early part of that revolution. It would be followed by the company's Super Bowl commercial in 1984 that launched Apple into popular culture. The sixty-second ad, inspired by George Orwell's novel *1984*, presented a repressed sci-fi future that declared—without naming names—war against IBM's drab vision for the future. "Apple Computer will introduce Macintosh," a voice intoned. "And you will see why 1984 won't be like *1984*."

*   *   *

By the time Tim Sweeney arrived at college to study mechanical engineering, he figured he knew all there was to know about computers. The thought of studying computer programming properly in school didn't interest him. Part of the issue was that he didn't see how to make money from programming free games shared through online bulletin board systems. And money was on his mind. His first job was at a hardware store making around $4 an hour, and he despised it. Then he turned to mowing lawns, a gig that ignited his entrepreneurial side. Soon, he was making $20 an hour but trying to figure out a way to make more.

He was toying with an idea involving computers, maybe a consulting business making customized databases. His preconceived notions about computer games changed, however, when he noticed a new business model emerging by the likes of Apogee, which became known for the PC hits *Duke Nukem* and *Wolfenstein 3D*. The shareware company began releasing a free taste of its games on the BBS, then charging for the full offering mailed to customers directly.

That inspired Sweeney to release *ZZT*, which he had been working on for several months at night. Orders came in as soon as he released it, and he began feeling flush as he raked in about $100 a day in sales. For a while, he continued living at his dorm while operating the business out of his parents' home, where he would copy the software on disks and package them up for mailing. Soon, he was working around the clock between school and coding his next game, *Jill of the Jungle*, which he launched a year later. Along with the new game, he changed the company's name to Epic MegaGames to sound like a legitimate competitor to Apogee, a fact he'd joke about later, calling the name "kind of a scam to make it look like we were a big company" and not just a guy working out of his parents' house.

His entrepreneurial side was excited. Sweeney was reading business magazines and studying accounting. Epic MegaGames was going to be a real company. And he was all in—dropping out of college one credit short of graduating and moving back in with his parents to run his nascent business. He had ambitions of being more than just a coder; he wanted to be a distributor for other game creators, promising to help market and deliver their offerings in exchange for a cut of sales.

One of the first people to reach out to him was a California high school student known on chat groups as CliffyB. Bleszinski, who would go on to success with *Gears of War*, began his career with a letter to Sweeney, including a game he had designed called *Dare to Dream*.

By happenstance, the package arrived at the same time an industry gaming executive named Mark Rein had come to visit Sweeney to talk him into hiring him as his first full-time employee. Together, Rein and Sweeney would make a formidable team in the cliché way: Rein had a gift for gab, while Sweeney was the technical guy. They both immediately saw potential in Bleszinski, and Rein phoned him with an over-the-top sales pitch. "You should publish this through Epic— you'll be a millionaire!"

Sweeney's business model was simple: Developers, guys like Bleszinski, would get 40 percent of revenue from the game sold, while Epic would keep 60 percent. To Sweeney, it was a good deal for developers because publishers such as himself tended to keep 70 to 85 percent. Sweeney was giving up potential revenue for the little guy. For Bleszinski, the initial royalty checks felt like a windfall. Though his first game didn't do so well, it did bring him into a growing fold of developers collected by Sweeney from around the world. For years, they worked remotely: Sweeney in Potomac, Rein in Toronto, Bleszinski in California, and so on. Sweeney kept an 800 number so they—mostly young people with little money—could call him for free. Others would just send him lengthy, long-distance phone bills they racked up talking with each other.

Bleszinski's next game, *Jazz Jackrabbit*, inspired by *Sonic the Hedgehog*-like games, would become Epic's biggest-selling game at the time. Another developer, James Schmalz, was so convinced he could have a career in gaming after making $50,000 in royalties with a game called *Solar Winds* that he began working on what would become another hit, called *Epic Pinball*.

"I went from earning $1,000 a month to earning, at times, almost a hundred times that," Schmalz told a gamer publication. The proceeds helped fund Epic's pivot from a successful shareware company to a major force in the video game industry.

In the late 1990s, video games were changing from two-dimensional characters, often bitmaps that sort of floated on the background, into almost lifelike, three-dimensional graphics. As the team was working on a new game that called for a dragon flying over a terrain, Sweeney was creating an editor tool and had a realization that like with *ZZT* he was really creating something bigger: a video game programming system called an engine. That engine would help make creating games easier, allowing the creatives, like Bleszinski, the freedom to build where their imaginations took them rather than being hampered by trying to figure out the right code as he did years earlier. That realization would put Epic on a dual path: Sweeney would be working to create a video game engine that would be dubbed Unreal, while his team would be working to use the tool to create bigger and better games.

The strategic change also marked a shift in business models for Epic: Instead of publishing others' games, Epic would make its own and license out its tools to developers. It was moving from shareware to bona fide game developer. Bleszinski and his cohorts would become employees in an ever-growing company. Instead of being spread around the world, they picked Cary, North Carolina, for a headquarters—not exactly central to the tech industry. But an appealing spot because of a lower cost of living than Silicon Valley while still being close to hiking and beaches.

There was little time for those perks, however. For Sweeney and many others, life was Epic. The young men worked long, late hours. And Sweeney, at times, never seemed to stop, going days at a time.[4] There were exterior signs of their success, to be sure. Epic's parking lot would be filled with Lamborghinis and Ferraris with license plates that read "EPICBOY" and "GRSOFWAR."[5] Sweeney's own car collection would grow impressively to include two Lambos, three Porsches, some Hummers, and a Corvette. "It turns out having fast cars is an excellent hobby when you're a workaholic because even if you don't have any free time, you can always drive the car to work," he told the documentary crew.

Except a flashy car collection wasn't enough for Sweeney. In time, he would mature, expressing regret for wasting his money. His ambitions would turn to building something more meaningful. He would be consumed with the idea of fairness. He would see the world of video games was in for another revolutionary change. New worlds to be explored and controlled.

"WITNESSING HISTORY"

2008

It was unlike most Apple events. Subdued. Boring. Business. Phil Schiller stood onstage to welcome an auditorium of roughly three hundred people—journalists, Apple employees, and Silicon Valley dignitaries. They all had come to the company's sprawling headquarters at 1 Infinite Loop in Cupertino, California, for the next development in the iPhone. They were gathered for arguably one of the company's most important announcements, though at the time most probably didn't realize just how world changing an event this would be. The somewhat unexciting venue suggested as much. A little more than a year earlier, Apple had rented out the Moscone Center, located about forty-five miles north in San Francisco, to reveal the iPhone to an audience of four thousand.

On this day in March, forty-seven-year-old Schiller's baby was being revealed: the App Store. It would serve as the digital marketplace for the iPhone, allowing third-party developers to offer their software for download onto Apple's smartphones through a Wi-Fi or cellular connection. It promised a world of change. No need for a software maker to find a distributor to get their goods to market. No need for a customer to take a trip to the mall to buy the software. No need for a disk to download that software onto a computer. It was a world not originally envisioned by Schiller's boss, Apple's chief executive, Steve Jobs. The very idea of an App Store was a big change from how the iPhone was introduced, but one, in hindsight, that would seem so obvious. Worried about security, Jobs had been against the idea of allowing third-party software on the iPhone. That's what the web was for, and the iPhone's three-and-a-half-inch touch screen offered a then-unheard-of web surfing experience. But if anyone could persuade Jobs to change his mind, it was Schiller.

After Jobs, no one, perhaps, was more familiar to the Apple faithful than Schiller at that point. He was seen as a possible successor for the company's iconic co-founder, whose failing health was of constant concern. Schiller was the company's top marketer, speaking publicly at product reveals and helping shape how Apple projected itself to the world. He was also behind the scenes helping get those products out, working closely with designers and engineers. Schiller was credited with coming up with the idea of a scroll wheel on the iPod.[1]

Physically and stylistically, Jobs and Schiller were opposites. Jobs was slim and stylish, Schiller heavier and fond of ill-fitting button-down shirts. Jobs was uninterested in sports; Schiller, a Boston native, was a rabid hockey fan. But when it came to product, they were so in sync that insiders dubbed Schiller Jobs's Mini-Me, a nod to the *Austin Powers* movie. He embraced the nickname, keeping a cutout of the character in his office.[2]

Schiller had grown up in Massachusetts, dreaming of becoming a marine biologist. He learned to scuba dive before learning to drive. One summer, as a teenager, he worked at the New England Aquarium and was asked if he wanted to volunteer for a special event with Jacques Cousteau, the famed oceanographer and filmmaker. He would remember that event as his single greatest workday. No. 2 was when Jobs, who had been ousted from Apple, returned in 1997.

Apple had been near bankruptcy when Jobs returned to run the show after being gone for a dozen years. At Jobs's request, Schiller, who had spent six years at Apple before leaving, rejoined, too. Together, they quickly narrowed the company's sprawling lineup and, by 2001, had a new product that would put Apple on a new course, the iPod.

The success of the digital music player helped Apple move beyond the computer market into personal electronics and also opened entirely new business models, the music industry. The iPod had been conceived in an era when digital music piracy was rampant. Apple's iPod allowed users to download songs from compact discs onto their Macs, then transfer the files to the portable player. More important, though, Apple figured out how to create a digital music store for downloading those songs onto the iPod.

Jobs convinced the record companies, which were seeing CD sales plummet from online piracy, that they should sell their songs digitally for download through Apple's desktop music store, iTunes. His store would offer piracy protections—he promised—while making buying new music easy. Under the deals that Jobs cut, the record companies would keep 70 percent of sales while Apple would pocket a 30 percent commission. While it might have seemed like just pennies at the beginning for Apple, the significance of the revenue split—70-30—would become foundational for Apple, living on in its ethos like the Ten Commandments. At the time, however, it was easily dismissed.

Some outside observers believed music would be a money loser for Apple, simply a ploy by the company to juice hardware sales. In truth, however, offering the bundling of a physical device (the iPod) and digital good (music) created a powerful marriage for Apple. It was a combo made even more power when Apple made the unusual decision—unusual for Apple, that is—to create Windows-compatible iPods after its first generation of devices. That allowed the iPod to operate within the entire world of PCs, opening Apple up to a much larger customer base than just those using the Mac. It was a feat probably made easier because Microsoft was under pressure to play nice as it defended itself against antitrust claims brought by the U.S. government and European Commission.

By early 2008, Apple was estimated to be the second-largest music retailer in the United States, and *Billboard* magazine calculated that iTunes sold more than 1.7 billion downloaded songs in the previous year. Using that volume estimate, *Billboard*'s back-of-the-envelope math suggested iTunes revenue totaled $1.9 billion in 2007, which, if correct, would have been about 8 percent of the company's total sales during its fiscal year.

Apple might have been a computer company in 1984, when it shocked the world with its Super Bowl ad, but by 2008 it had also become a music company. And Jobs was worried about what would happen if the logical next step occurred in personal gadgets: the merging of digital music players with smartphones. The iPhone was Apple's answer to that. It was three devices in one: a music player, web surfer, and phone.

On January 9, 2007, Jobs stood onstage at Moscone to reveal the iPhone. "Every once in a while, a revolutionary product comes along that changes everything," he began. It was the sort of bravado that Jobs was known for, but in a short order of time he would be proven correct. As exciting as the iPhone promised to be, it had an obvious hole to techies outside Apple: its strategy for offering third-party software. Almost immediately, outside developers were clamoring for the ability to create software to run natively on the iPhone's operating system, dubbed iOS. Nitin Ganatra, director of engineering of iOS applications, remembered hearing the requests. "Hey, you know, web technologies aren't . . . going to work as well as what you demoed, so why is that the answer for third parties?" he recalled hearing from developers. "And, you know, I think more and more over time, those questions must have just worn away at Steve and other people."[3]

Inside Apple, software engineers were already developing more apps to take advantage of the iPhone's hardware, such as an app to play YouTube. "Why?" Ganatra recalled years later. "Because if they went and used web technologies, it would have been a much different user experience. It wouldn't have been consistent with everything else on the phone." In other words, it would have sucked. Scrolling would have been weird, for example. And it wouldn't have run as smoothly.

But more than just a cooler user experience, executives and engineers began thinking more and more about the marketplace for the iPhone and how it could be used as a business tool. At the time, BlackBerrys were must-have business accessories. Its rival handset maker Palm had a program popular within the medical profession that ran third-party software called Epocrates that allowed doctors to quickly pull out their device and look up medical information. Apple began wondering how it could win over doctors, who were eager to use iPhones. Either Epocrates would have to figure out a web-based program that would truck a ton of data to the iPhone so the info was available instantly, or the program's developers would need a persistent web connection to make their software work. All of which sounded different from the work it had already done for its successful Palm application.

Even if Apple could persuade Epocrates to do that, whatever was created for the iPhone would then be lesser to Apple's own app experience. "Very quickly we realized that . . . in order for that company . . . to create something like Epocrates for iPhone they would have to do an enormous amount of work," Ganatra said.

Plus, the whole idea of Apple having a different process from everyone else flew in the face of the company's philosophy. "Apple hates it when we have technology that is being used internally that is different from what's being used externally," Ganatra said.

Some on the team suspected that Jobs's reluctance to allow native apps was more than just tied to security concerns. He might not have trusted outsiders to offer an app that fit with Apple's aesthetic. Would they, for example, know how to design for a touch screen?

In the end, however, the best argument to Jobs might have lain in the user experience.

With web-based programs, Apple wouldn't be able to control the user experience in the same way as with native apps. Jay Rubin, a program manager who was originally brought onto the project to help with the web programs, remembers hearing from project managers pleading for help to improve things. "This is a horrible experience," they kept telling him. "We're not getting performance. The phones are chugging down. There's no way to control."[4]

Soon enough, Jobs was convinced. The future of the iPhone would be apps. Companies such as Salesforce and Sega were brought in early to try to develop their own apps ahead of the software kit being released to the public. Their experiences—and apps—would be part of the rollout as real-world examples for how easy and useful the App Store could be.

Around the same time, Apple was working with coders at Yahoo to integrate the web company's email service into the phone. Yahoo had an idea that was a step further than Apple's idea of apps; it wanted to use its own app software on the iPhone, presumably as part of a larger effort to spread software widgets across platforms. Schiller disliked the idea and reminded his colleagues to remain true to their strategy. Apple had been developing its own application programming interface, and if it began letting Yahoo use its own, then why not let Sun, Adobe, or Microsoft? Even worse, he warned, Apple would lose control. "With one API (ours) we can manage what is on our products and what is not," Schiller wrote a colleague in a memo. "If we open it up then we don't sign all apps, we don't distribute all apps, etc. Which is the same as throwing out the whole plan we have in place." It was a hard no.

It was still early days, but inside Apple they were already talking about how the ecosystem of products could expand beyond the iPhone to what we would know as the iPad and Watch, and the App Store was seen as a way to tie those pieces of hardware together, Rubin said. And it was Schiller who saw the potential to help Apple's bottom line. "He was always a couple of steps ahead of you," he said. "I don't think he wanted to just immediately jump on the team to say, 'Hey, do this X, Y, and Z.' He had a completely different approach than Jobs did. Where Jobs was, 'I want you to do X, Y, and Z,' and not even tell you why, Schiller was more of, 'I think this is around the horizon, what are your thoughts?'"

When it came to revealing details of the App Store itself, Jobs took to the stage on that day in March. He wasn't talking to customers; he was talking to Silicon Valley developers, those whose sweat and blood would fuel this new world that Apple was proposing, whose software dreams would make the iPhone more than just a way to surf the web or send an email or make a phone call. "We think we've got a great business deal for our developers," said Jobs, gaunt from fighting cancer and dressed in his trademark black half turtleneck.

Then, harking back to the days of iTunes, he announced that developers would be able to pick the price for offering their apps. "When we sell the app through the App Store, the developer gets 70 percent of the revenues right off the top," Jobs continued. "We keep 30 to pay for running the App Store. . . . This is the best deal going to distribute applications to mobile platforms."

The 70/30 split would only apply to digital goods. In the years to follow, Apple wouldn't apply the commission to transactions occurring within the app for physical goods—such as shoes purchased through the Nike app or rides ordered through the Uber app. The difference being that digital goods were being consumed within the Apple ecosystem whereas shoes weren't. Nike was having to mail sneakers to a customer's home—Apple wasn't getting involved in all of the messy details of physical life, such as taxes and lost deliveries.

Developers that want to give away their apps for free would face no charge from Apple, he said, drawing applause from the audience. Apple wasn't going to pass along credit card fees or marketing fees. There would be limitations, he said. No porn or malicious apps or ones that invaded user privacy. "So there will be some apps that we're going to say no to," Jobs said. "But again, we have exactly the same interest as the vast majority of our developers, which is to get a ton of apps out there for the iPhone and we think we've invented an incredibly great way to do that."

As he was known to do, Jobs didn't just end with that. "We do, though, have one," he said to laughter, "one last thing." He presented a stunning endorsement of Apple's vision for mobile software: John Doerr, the head of the high-profile venture capital firm Kleiner Perkins Caufield & Byers, was on hand to announce it was creating a new fund to invest in developers making iPhone apps. "The computer scientist and Apple fellow Alan Kay is famous for saying the best way to predict the future is to invest in it," Doerr said. "And at KPCB we like to say the second-best way is to fund it." To underscore how big of a bet Doerr was willing to make, he noted that it had taken a couple million dollars to start the video game behemoth Electronic Arts, $8 million to start Amazon, and $24 million to create Google. For Apple's digital world, Doerr was willing to bet much more: *$100 million*. "Today we're witnessing history," he said. "Think about it: What the iPhone is all about is in your pocket. You have something that's broadband and connected all the time, it's personal, it knows who you are and where you are. That's a big deal—a really big deal. It's bigger than the personal computer."

And Apple was offering up keys to that new world.

*   *   *

A month after launching the App Store, Steve Jobs summoned *Wall Street Journal* reporter Nick Wingfield to Apple headquarters.[5] Jobs was excited. After launching with five hundred apps, the App Store now had an inventory of more than fifteen hundred. About 30 percent of them were being offered for free. Most of the paid apps were being sold for less than $10. And iPhone users were eating it up.

"Users have downloaded over 60 million apps from the App Store in the first 30 days," Jobs told the reporter. "That is 30% as big as iTunes for music downloads.

"Let me say that again," Jobs said for effect. "App downloads equal 30% of all iTunes song downloads during the last 30 days."

"What does that number say to you?" Wingfield asked.

"It says the App Store is much larger than we ever imagined," Jobs said. "iTunes has been out for over five years. In 30 days, users downloaded 30% as many apps as everybody in the world downloaded songs from iTunes."

As the reporter tried to ask another question, Jobs talked over him to drive home the point of just how thrilled the company was at the gamble it had taken. "We didn't expect it to be this big," Jobs said. "The mobile industry's never seen anything like this. To be honest, neither has the computer industry."

Jobs laughed. And again he said the numbers as if even he couldn't believe them. "Sixty million downloaded applications in the first 30 days," Jobs reiterated. "Thirty percent as big as iTunes song downloads during the last 30 days. This is off the charts."

The App Store came just as the nation's economy was to be rocked by the Great Recession—a downturn in real estate, subsequent collapse of Lehman Brothers, and eventual bankruptcies of General Motors and Chrysler—that left many jittery about their own finances and futures. Suddenly, if you knew how to code and had a quirky idea, there was an outlet for trying to make a quick buck. And one of the most popular apps to emerge quickly helped cement the App Store as a place for lone software developers to try their luck at being entrepreneurs.

Steve Sheraton, a broke magician living on his friend's couch, would become proof of that possibility.

The thirty-seven-year-old liked a good visual gag. Soon after the iPhone had come out, he had created a looping video of himself drinking a beer for YouTube that gave a pretty good illusion of a beverage being consumed on the iPhone's colorful screen.[6] With the App Store, he became an early example of a developer creating something that Apple never anticipated. Being a native app would give his idea new pop. The entire screen could look like a beer glass, and the magic of the iPhone could make it look as if a user were drinking that beer.

For Apple, his app would help illustrate just one of the benefits of being able to access iPhone's hardware that wasn't available to web-based apps: the phone's built-in accelerometer, which measured the motion of the phone. Sheraton could use the sensor to measure the phone's angle versus the horizon. His app, dubbed iBeer, would then tether the line between his image of "liquid" and "foam" to the horizon. By doing so, when the phone moved in a certain direction, it looked as if the liquid were moving as well. If the line went past a certain spot, then the image would look as if the "glass" were getting empty. Shaking it would give the appearance of foam.

It was a fun party trick. And soon it became *the* party trick—the perfect way for proud new owners of iPhones to show off their new gadgets. Almost immediately, iBeer ranked No. 1 on Apple's paid-app list. Sheraton charged $2.99 for iBeer, available only for iPhone users through the App Store. Apple kept 30 percent, or almost 90 cents per purchase. For Sheraton, it was life-changing. "The amount of money that was coming in was just so over-the-top," Sheraton recalled years later. "During our heyday, we were making $10,000 to $20,000 a day."[7]

Other developers took note, but soon learned that bringing out a hit app wasn't as simple when Apple wasn't asking you to create it, even if Apple was excited to have the unexpected wave of developers submitting their ideas for review to be allowed in the store.

Phillip Shoemaker had already quietly purchased an iPhone, careful not to let his bosses see. He worked at a startup in Silicon Valley working on artificial intelligence software created by one of the cofounders of Palm. He had followed him from Palm, where Shoemaker had worked on a project that attempted to create an app store for its devices.

After Apple announced its App Store, Shoemaker quit his job and began trying to create his own apps, tinkering with his son on ideas, including one with fart sounds. Shoemaker's excitement was soon tempered as his submissions to the App Store were rejected. He'd make a fix, resubmit, and wait to hear back. Three weeks later, the reviewer would reject it again. He would make the new adjustment, then resubmit and wait again. And on and on it went.

He soon realized the review team wasn't giving his app a full review and rather was stopping at the first issue that arose. "It was like reviewing someone's résumé where you stop on the first typo, give it back to them, and then you go on to the second line," Shoemaker said. "That would drive a developer crazy."

It was driving him crazy. He began writing emails to complain, sending them to nameless email addresses at Apple, like throwing pennies into a deep well hoping someone might grant a wish. Email after email, long tomes detailing what he saw as the problem with the app review process, what should be done differently, how irresponsible Apple was being. He went on and on and on.

Eventually, he got a call from Apple. An executive named Eddy Cue wanted to meet with him at Apple headquarters. Shoemaker wasn't an Apple fanboy. He had bought some Apple stock after it came out with the iPod years earlier, impressed with the vision for digital music. He loved his iPhone and believed in the potential for the App Store. But he didn't know that Cue was one of the most senior executives at the company, the man responsible for the infrastructure of the App Store. He accepted with a sort of naive hubris. He'd go in there, he thought to himself, and maybe, finally, somebody would listen and the process could get better. And iFart would be unleashed on the world.

When the day came in late 2008, Cue sat down with him in the lobby of the company's headquarters. "You've written a lot of emails, you've submitted some apps, I've seen all your apps," Cue told him.

Not surprisingly, Cue didn't think the fart app was going to change the world. But, Cue told him, Apple knew it needed to improve its App Store process and was creating a position to oversee the review process. Cue wanted Shoemaker to apply. What would follow over the next few months was a dizzying number of interviews, including with the marketing chief, Phil Schiller, that culminated with a meeting in the company's boardroom.

Shoemaker found himself seated on one side of a long table facing Jobs, who was flanked on his left and right by senior executives, including Cue and Schiller. On the table directly in front of Shoemaker sat several iPhones for the final test. A pop quiz of sorts.

"We want you to review some apps in front of us," Jobs said.

"What are the guidelines?" Shoemaker asked.

"There are no guidelines," Jobs told him. "You tell us."

Shoemaker took the first steps of what would become a yearslong journey in defining how Apple would police its App Store. On that day, he went through each app, saying what he thought. Was this allowable or not? This looked offensive, but was it okay? He had more questions than answers because there were no set rules. And that was the problem. That would be the job, ultimately, to codify Jobs and Schiller's views of what was okay and what wasn't. "They wanted to hear how I responded, how I debated, how I argued," Shoemaker said. "Steve would say 'No, what about this?' and I would argue with him, and Phil would argue with me. And I would just give it back as good as I got because, honestly, I had nothing to lose. I didn't really want a job. I was having fun writing iPhone apps at home and playing with my kids."

But sitting at the table, bantering back and forth, was fun, and soon it became clear he was the man for the job.

"MORAL
RESPONSIBILITY"

2009

The call came from Steve Jobs's office. The CEO wanted to talk. *Now.*
Three weeks into the job at Apple, Phillip Shoemaker was getting
a sharp introduction to what was at stake overseeing the company's
nascent app review team. He had six employees. App submissions were
piling up. Unbeknownst to him, his team—if you could even call it
that—had approved a new game called *Baby Shaker*. The 99-cent app
was, as the name implied, a game where users shook the phone to
kill an image of a cute baby. Its release the day before had quickly
gained attention—outrage, really. *The New York Times*, *The Washington
Post*, and others wrote about it. The company's stock dropped. Protest-
ers were picketing outside Apple's New York City store. "What have
you done?" one of the company's public relations officials had already
called to ask.

Shoemaker didn't know. He was still going through training. His
team didn't even have an official office, squatting in glass conference
rooms. He hadn't even seen the app before it was approved and posted
to the App Store by another manager. But as soon as Shoemaker saw
the app, the father of two young children knew the gravity of the

situation. This was bad. And now the CEO was calling. Jobs kept it brief: "You are stupid, and you hire stupid people."

Click.

A panic hit Shoemaker. He rushed to the office of his superior, Phil Schiller, to ask, *Am I fired?*

"That's just Steve," Schiller reassured him. But it also wasn't a definitive assurance that Shoemaker would, in fact, keep his job. Time would tell. A year after announcing the App Store, Apple was getting a bitter taste of what it could be like exercising editorial judgment. It had opened itself up to a new level of complexity and unintended consequences. The company was also trying to get its arms around unexpected success and the natural competition it was attracting in the smartphone business, including from its archrival, Microsoft, and longtime friend Google, which was trying to branch out from the online search business with its own mobile ecosystem known as Android. A new Apple ad campaign aimed at highlighting the advantage the iPhone had with its newly preinstalled App Store that came on every phone. It introduced a catchy tagline that would enter mainstream parlance: "There's an app for that."

"What's great about the iPhone is if you want to check snow conditions on the mountain, *there's an app for that*," the commercial began. "If you want to check how many calories are in your lunch, *there's an app for that*. And if you want to check where exactly you parked the car, *there's even an app for that*. Yep, there's an app for just about anything. Only on the iPhone."

But Apple wasn't just trying to distinguish its iPhone as home to a colorful assortment of useful apps, it was also trying to establish itself as a safe world away from the ugliness of the internet at large. In many ways, it carried similar shades of how Walmart then refused to sell music CDs carrying the Parental Advisory tag. For Jobs, there was zero tolerance for porn in the App Store. The iPhone Developer Program License Agreement seemingly banned such content—except implementing the legalese was proving controversial and difficult for Shoemaker's team.

In another particularly embarrassing episode, Shoemaker's team rejected an app from the editorial cartoonist Mark Fiore, who won the

Pulitzer Prize in 2010, because the company told him his app contained "content that ridicules public figures and is in violation of Section 3.3.14 from the iPhone Developer Program License Agreement which states: Applications may be rejected if they contain content or materials of any kind (text, graphics, images, photographs, sounds, etc.) that in Apple's reasonable judgment may be found objectionable, for example, materials that may be considered obscene, pornographic, or defamatory."[1]

Once the rejection became public, Jobs would call the Fiore episode a mistake. It raised questions about why some apps were getting rejected while others weren't. *Sports Illustrated* and *Playboy*, for example, were allowed pictures of scantily clad women, but others weren't. Some even questioned why porn was banned. That was a nonstarter for Jobs, who defended Apple's effort to keep porn off the app, emailing a reporter a terse defense: "We do believe we have a moral responsibility to keep porn off the iPhone."

He added, "Folks who want porn can buy [an] Android phone."[2]

*   *   *

Apple hadn't invented the idea of an app store, but its execution of the iPhone hardware had supercharged the new approach to how software could be sold and disrupted. It came roughly a decade after the dotcom bubble emerged around the idea of the consumer internet, injecting hundreds of millions of dollars into new businesses imagining how commerce could become digital. By 2009, one of the clear winners was Google and its unique algorithms that unlocked how non-techies navigated that new digital world. Users searching for information through Google in exchange gave the company vast amounts of information that in turn allowed its founders, Larry Page and Sergey Brin, to develop a juggernaut of an advertising business. That success had given them a war chest to pursue other bets. In 2004, Google acquired a startup that would become Google Maps. In 2005, it gambled on another startup called Android that would become even bigger than turn-by-turn direction.

Founded in 2003 by Andy Rubin, a distant relative of Apple's Jay

Rubin, who would work on the App Store, Android was initially envisioned as a creator of software to run digital cameras, which were growing in popularity. He and a cohort of engineers, however, quickly realized few investors shared their vision. Instead, the crew was encouraged by Silicon Valley investors to lean into their experiences in early cell phones to create an operating system for smartphones. Rubin had been one of the founders of a startup called Danger that had gained attention for its work creating a smartphone called the Sidekick. The phone had been a pioneer in integrating users with the internet—including using Google as its default search engine.

Even if Rubin had been reluctant to return to smartphones, the business case for creating an operating system was compelling. Even before the iPhone, it was clear to some that mobile phones were on a pathway to essentially becoming pocket computers. The amount of computing power found in typical devices in the early '00s was similar to what was found in desktop computers just a few years earlier. All of that computing power was just crying out to be tapped. And the cost of making the handsets was only getting cheaper. Yet the challenge lay with the software to make the hardware work.[3] It was still expensive, and there were few options for handset makers. Microsoft, the dominant player in desktops, made money licensing its operating system to handset makers. BlackBerry's Research in Motion had a very Apple-like setup with its own software for its hardware that it didn't share with rivals. And the other options that existed weren't great, meaning third-party software made for one manufacturer's phone wouldn't work on another maker's device.

Rubin began pitching Silicon Valley investors on the idea of an open-sourced operating system that handset manufacturers could use for free. In turn, the Android team imagined a large ecosystem of phones running its operating system that would then entice software developers to create apps that could run across a large base of devices, similar to what happened in the PC except instead of paying for Windows. The idea appealed to Page, who had been a fan of the Sidekick. Instead of investing, though, he proposed Google acquire the nascent startup with the promise of funding its work to build out its designs. They closed a deal in 2005—for an estimated $50 million—and Rubin's team began racing to create their digital world, a world that became more appealing to others when the iPhone was revealed.

The iPhone's debut was a gut punch to Rubin's team. At the time, they had been working on software for a touch-screen phone by HTC but had hoped that their operating system, by then called Android, would be ready for a text-based phone first. That was scrapped as they aimed to compete with Apple with the HTC Dream phone in 2008. Their first try got lukewarm reviews. But it showed what was possible. By then, the industry was looking for ways to compete with Apple. Gadget makers and cellular service carriers feared they would be left behind by Apple's hit new product and what it might grow into. And an open-source model was appealing. "If you have a company like HTC and they're getting charged $10/unit by Microsoft, and they have to do a whole crazy bunch of integration work with Microsoft in order to get it to work, the idea of free and open source is magic," Michael Morrissey, who had joined the Android team from Microsoft, later recalled. "If you have Android, then your [phone maker] can bring up new devices super, super fast, because they have access to all the code. On top of that, it's free."[4]

Months after Apple had first announced its App Store, Android revealed its own plans. "Developers will be able to make their content available on an open service hosted by Google that features a feedback and rating system similar to YouTube," Eric Chu, the head of Android's developer ecosystem, told developers in a blog post. "We chose the term 'market' rather than 'store' because we feel that developers should have an open and unobstructed environment to make their content available. Similar to YouTube, content can debut in the marketplace after only three simple steps: register as a merchant, upload and describe your content and publish it."[5] While Google was leaning into the term "market" rather than "store," the mechanics of the scheme were very similar. Those software developers that charged a fee would keep 70 percent. Unlike Apple, however, Google said at the time that it wasn't keeping the remainder of the fee for itself. Instead, the company said, "the remaining amount goes to carriers and billing settlement fees. . . . We believe this revenue model creates a fair and positive experience for users, developers, and carriers."[6] That would change over time to the dismay of developers, igniting the broader pushback against Apple when the world fell into two categories: Apple and Android trying to be like Apple.

But in 2008 and 2009, Google was still racing to catch up, building a world that would ultimately look similar to Apple's ecosystem but structurally had its own challenges. By its very nature, Android would have to run on a wide variety of devices while Apple could tailor it for its own hardware, tuning it to run optimally. Google would also have a large and diverse network of partners to appease. Privately, Google was making its case to phone makers for why its app store was the best strategy and why each manufacturer shouldn't follow suit with its own version. "We certainly believe that a single 'app store' is an essential piece of the strategy to make the overall Android ecosystem successful, and we are putting significant investment into making Android Market that single *open* distribution system," a Google executive wrote to Samsung in a 2009 memo. "Google operates Android Market as a revenue-neutral service—we do not seek to profit off of application sales, and we invest in Market because it is essential to the open ecosystem."

Apple watched Google warily. Internally, executives had dismissed what Microsoft was doing with mobile, and they concluded the company's inability to properly do hardware was making for a bad user experience. But Google seemed different. "The big threat from Google was that Google knew software and services," Bob Borchers, one of Schiller's top deputies as senior director of iPhone product marketing, later said. "In fact, probably knew them better than Apple did." To Apple, he added, Google felt as if it could "legitimately be a significant threat to a new platform like iOS."[7]

*   *   *

After the controversy over the *Baby Shaker* app, Phillip Shoemaker began each day with a check-in with Apple's public relations office so he could alert the department to any possible time bombs in the making. Even as Apple was trying to make the iPhone the place for developers, it was seeing a very public revolt of those unhappy with its seemingly inconsistent and unclear rules for what was okay.

One developer reported getting rejected for having an icon that looked like a small iPhone.[8] The Nine Inch Nails front man, Trent Reznor, said his app was apparently rejected because Apple objected to content on the band's album *The Downward Spiral* (it was eventually approved).[9] Another developer claimed his app—designed to educate users on the benefits of the single-payer health-care system—was rejected because of its politics.[10] Another app called *Zombie School* was removed after complaints that a game shooting zombie students on an infected campus promoted school violence—a claim the creator, Retarded Arts, denied.[11] John Gruber, a popular blogger who closely followed Apple, was uncharacteristically critical of the company when he accused it of forcing *Ninjawords*, a dictionary app sold for $2, to censor itself in order to get on the App Store. Omitted words included "ass," "snatch," "pussy," and "cock." "Every time I think I've seen the most outrageous App Store rejection, I'm soon proven wrong. I can't imagine what it will take to top this one," Gruber concluded. "Apple requires you to be 17 years or older to purchase a censored dictionary that omits half the words Steve Jobs uses every day."[12]

After reading the blog, Schiller dug in and tried to defend the decision, saying *Ninjawords* provided access to more vulgar terms than found in traditional dictionaries and said Apple hadn't asked the developer to censor the words. Rather, the Apple app reviewer had merely suggested resubmitting the app for approval once parental controls were implemented on the iPhone—a new feature being rolled out later that year. Instead of waiting, the developer censored itself.[13]

Even Shoemaker was drawing critical attention from the media. "Apple's App Store Director Sells His Own Fart Apps," an article on *Wired* magazine's website read. "Apple has long been an icon for quality products, but its overflowing iOS App Store is a crapshoot: Nuggets of quality are buried in a vast, steaming heap of inanity," Brian X. Chen wrote. "In fact, the man who oversees the App Store process runs a side business selling fart and urination apps."

Shoemaker's apps, *Animal Farts* and *iWiz*, which originally helped him understand what was wrong with the App Store, were now the cause of a new PR nightmare for Apple. It didn't matter that his apps were created and submitted before he was hired. From the outside, Apple appeared to be floundering with how to manage the App Store. "Google has done a better job at leveling the playing field for independent developers, and that matters," Jeffrey S. Hammond, an analyst at Forrester Research, told *The New York Times*. "I don't think Apple can tolerate that, especially with the dozen or more pads that we're going to see hit the market in eight to nine weeks."[14]

The introduction of the iPad tablet in early 2010 also complicated things. Apple had been courting media companies to create content for the new device, further expanding the role of the App Store plus adding a new avenue for commerce, through its iBooks Store app.

As the company moved further into being a gatekeeper for content, Apple was entering new territory. And early brushes with controversial content would look prophetic when it came to how the company's leaders might rule when a matter disagreed with the liberal views commonly held in Silicon Valley. Shortly after introducing a digital book app, for example, a publisher wanted to launch a biology textbook that intended to put creationism side by side with evolution. "I normally don't get involved in politics, especially within the company, but . . . this was just . . . against everything I believe," Richard Williamson, a director involved with iOS software engineering, recalled years later. He remembered telling his boss's boss, Scott Forstall, head of software, "Look, this is totally not acceptable. We need to do something about this."

Surprising to some, however, one of the company's most senior executives was for including the objectionable content. "Turns out Bob Mansfield is a very religious guy," Williamson said, "and he really wanted to include this textbook."

So heated did the internal debate grow that it eventually found its way to the company's board of directors. Ultimately, a decision was made to prohibit allowing a side-by-side comparison. Or, in other words, Apple put its thumb on the scale for a certain worldview—one against some conservatives. "That's the only thing in terms of censorship that I've ever really come close to dealing with at Apple," Williamson said.[15]

If only things had been so easy for Shoemaker.

He had been attracted to the job of overseeing the App Store review process because of that experience as a developer waiting months to have his apps go through. And he was intent on speeding things up—holding a dream in his head for one-day turnaround. He proposed to automate the process, but Jobs balked. Jobs wanted human eyes to have the final sign-off on every app. In those early days, the team was rejecting most apps—about 70 percent. Those who were rejected would often just turn around and reapply, sometimes without any clear idea why they had been booted to begin with. It might take an app three weeks to get reviewed, even with reviewers eager to help.

Each week, Shoemaker would take the trickiest apps for consideration to the Executive Review Board, which consisted of some of the company's most senior executives. There was Schiller, whom they began calling the chairman of the App Store, because of his intent focus on its operations; Eddy Cue, who oversaw the company's digital services, including iTunes; and Forstall, the chief software engineer whose teams were responsible for making the iPhone hum.

Shoemaker was also racing to hire a staff, quickly discovering that the best place to pluck talent was from the Genius Bars at Apple retail stores. These mall employees were often Apple fanboys who pined for a job at corporate. More important, they had a sense of the Apple brand. Within months, the team would balloon. Still without an official workspace, Shoemaker was stuffing them in conference rooms where they would black out the glass windows and install key cards for entrance. It was all top secret. They were deciding how and why to reject or approve apps. And there was no official guidebook. At the core, they had the Developer Program License Agreement agreed to by all developers using the platform, which in legalese laid out some broad strokes but put nothing in plain English, leaving lots up to interpretation.

Instead, the team would try to interpret Schiller and the rest of the Executive Review Board's intentions from previous tough calls. For example, shortly after the *Baby Shaker* incident, Shoemaker told the team to reject any proposed app that included something "cute or cuddly" that dies. And like that, it was etched in stone, a rule passed on to new recruits. For years afterward, the team would reject anything that had something cute that died. A year later, a Canadian developer kept submitting a game that involved clubbing a baby seal, only to be rejected. "Finally, I got on a phone call with him," Shoemaker said. "He's like, 'Sir, you don't understand, this is what we do in Newfoundland.' And I was like, 'Hey, I'm sorry, that's done in Newfoundland, but this isn't something we allow in Cupertino.'

"If you're cute and furry or a baby," Shoemaker continued, "we're not going to allow you to die." Years later, Shoemaker would realize his team was rejecting the popular video game *Duck Hunt* because of the *Baby Shaker* precedent—even though it was a very different scenario. The team was terrified of getting fired for approving the wrong app, Shoemaker said. "We never fired anyone for approving the wrong app, but people thought that's what was going to happen."

Different scenarios and the subsequent decisions were logged in an internal wiki that eventually grew quite large. And for the longest time, Shoemaker would give final approval for an app to go live in the store. After a normal workday, he would spend the evening going through the queue of apps awaiting final approval, pushing the "approved" button. His goal was to get through five hundred a night. As he did that, he was seeing the lack of consistency between reviewers on what was approved—evidence in his mind that their policies weren't clear on what was okay. He lobbied Schiller to create official guidelines. But Jobs balked. *Too soon*, he would say. Jobs worried developers would begin creating to the guidelines, looking for ways to get around the intent and, in the process, creating more bad PR. Jobs did not want any more bad PR. His caution was understandable: Apple was riding a moment. The fiscal year 2010 was shaping up to be a blockbuster for Apple with iPhone unit sales looking to reach nearly forty million, almost double compared with 2009 and a 245 percent increase from 2008. The company's annual profit was on track to be four times what it was in 2007, when the iPhone launched. And the rocket ship looked to keep going higher, if they could just avoid making a mistake.

Every week Shoemaker would try to pare back the list of apps to be considered by the review board with pre-meetings, but still the official sessions would stretch on for four hours. These were apps on the bubble—not obviously ones that should be rejected, but also not clear cases for approval. Topics such as sexual content. Going back to that March 2008 App Store debut, Jobs had been clear that porn wouldn't be allowed. Still, entrepreneurial developers looked at the offerings on the App Store and quickly saw an opening for risqué offerings to try their luck: Apps that featured women in bikinis? Not bikinis but pasties? Bikinis showing nipples through the fabric? Bikini bottoms showing "camel toe"? What about men in thongs? Tight thongs, thongs like sausage wrappers?

Some of their calls seemed obvious, only for decisions to come back and bite them months later when the team rejected a comic version of the literary masterpiece *Ulysses* because it contained at least one scene of nudity. "I don't think the Apple representative that I first spoke with even knew what Ulysses was," the creator's business manager told a reporter.[16] Apple backtracked on the censorship after it became public.

A year into the job, Shoemaker was still trying to figure out how to speed up the review process. By then, the team was approaching two hundred people and had its own office building in Sunnyvale. He conducted motion studies to better understand why it was taking so long to review a new app. To him, the answer was clear: If his own team didn't fully understand what was okay, developers were even more uncertain. The rejection numbers would finally help Shoemaker make the case for putting out official rules. One day Schiller asked why 70 percent of the team's rejections were because of "objectionable" material? *Porn*, Shoemaker told him. Schiller didn't believe it. He instructed Shoemaker to review the results personally. For a week, Shoemaker instructed his team to forward any app rejected for objectionable material to him. "My email got flooded with pornographic pictures," he said. "To the point that HR even stepped in and said what's going on?" Every single day, developers were submitting new porn apps—not just one but hundreds per day. Eventually, Jobs and Schiller decided it was time to release the guidelines to help developers understand the review process. After months of work, they were published in September 2010:

We view Apps different than books or songs, which we do not curate. If you want to criticize a religion, write a book. If you want to describe sex, write a book or a song, or create a medical app. It can get complicated, but we have decided to not allow certain kinds of content in the App Store. It may help to keep some of our broader themes in mind:

- We have lots of kids downloading lots of apps, and parental controls don't work unless the parents set them up (many don't). So know that we're keeping an eye out for the kids.

- We have over 250,000 apps in the App Store. We don't need any more Fart apps. If your app doesn't do something useful or provide some form of lasting entertainment, it may not be accepted.

- If your App looks like it was cobbled together in a few days, or you're trying to get your first practice App into the store to impress your friends, please brace yourself for rejection. We have lots of serious developers who don't want their quality Apps to be surrounded by amateur hour.

- We will reject Apps for any content or behavior that we believe is over the line. What line, you ask? Well, as a Supreme Court Justice once said, "I'll know it when I see it." And we think that you will also know it when you cross it.

- If your app is rejected, we have a Review Board that you can appeal to. If you run to the press and trash us, it never helps.

In other words, Apple was handing down the rules for its kingdom. Other tech companies had an opportunity to operate within its walls, but they had to follow its rules—rules it could increasingly enforce because there was so much money to be made that developers didn't seem to care, at least initially.

4

A BOOK TOO FAR

2010

Eddy Cue, one of Steve Jobs's top deputies, seemed confident his secret mission to New York City was going to pay off. He was in Manhattan to visit with the most powerful people in the U.S. book publishing industry. The so-called Big Six: Random House, Penguin, HarperCollins, Simon & Schuster, Hachette, and Macmillan. After meeting with just three of the publishers, Cue dashed off an optimistic email to his boss back in Cupertino. "Nothing scared me or made me feel like we can't get these deals done right away," he told Jobs that evening.

If he was right, Cue and Jobs would soon be expanding Apple's growing Walled Garden to yet another realm, books, which he hoped would be as big a business for the tech company as music had become, if not bigger. Cue's mission would also put Apple on a path that would open it to claims of abusing its business success, creating an unlevel playing field that harmed consumers—claims it would deny but that would haunt it for years to come. But on that evening in early December 2009, Apple was greeted as a savior. "Everyone was ecstatic to see Apple and what it could mean for their industry," Cue told Jobs. Meetings were scheduled with three more publishers the following day; in total, the six publishing houses accounted for more than 60 percent of book sales. The bad guy to those book publishers was Amazon.com, which was sending shock waves through the market with Kindle, its digital book reader. For the book executives, Apple and its ambitions to compete in the e-book market seemed like just the thing that might give them leverage against Amazon.

*   *   *

Steve Jobs lacked interest in using the iPhone as a digital book reader. The screen was too small, he argued. Eddy Cue, who had built iTunes and worked with Phil Schiller to build the App Store, saw great potential. His team calculated that the North American book market was larger than the music business. They estimated the industry was as big as $42 billion with so-called trade books, those aimed at general readers, making up $12.5 billion. The e-book market, while dominated by Amazon, was still in its infancy with just $100 million in sales and growing exponentially. In 2010, Cue's team predicted the e-book market might reach almost $1 billion.

Given Jobs's objections, Cue eventually suggested they approach Amazon about agreeing to carve up the media market; Amazon would get e-books, and Apple would own digital music. They would cross-sell via their digital platforms. "I can't see them agreeing to this but if they really value books and want to own the category going forward than [sic] maybe they would consider it," Cue told Jobs in an email in early 2009. "At this point, it would be very easy for us to compete and I think trounce Amazon by opening up our own ebook store. The book publishers would do almost anything for us to get into the ebook business." While his idea didn't go anywhere, the proposal gave insight into how leaders in Cupertino saw the new digital world they were creating, a world to be cut up and controlled.

As the year came to an end, however, Cue would have another chance to pitch Apple's entrance into the digital book market. Work was underway to bring out the iPad in early 2010. The tablet computer was essentially a giant iPhone, which held the promise of doing everything the small-screen smartphone could do but also act like a laptop—but with a touch screen instead of a keyboard. Once Cue had a chance to touch and play with an early version of the iPad, he later said, he immediately became convinced Apple had a huge opportunity to become a serious player against Amazon. He raised the idea of e-books to Jobs yet again. This time, Jobs came around to the idea. "That was the good part," Cue said. "The bad part was that this was in November and . . . we were launching in January."

It was then a scramble. They were racing against time in more than one way. As Jobs was preparing the next step in Apple's transformation, he was fighting to stay alive. He had fought cancer, a battle that his body was losing. Cue could see the effects on his friend, mentor, boss, whom he had worked so closely with for almost two decades. Jobs was gaunt and pale. He had taken a leave of absence and received a liver transplant, but was back now, pushing for the next big thing. Jobs wanted to reveal the device at one of his splashy events on January 27, then begin actually selling the iPad in April. They wanted to quickly line up book publishers to tout at the event, helping demonstrate how the technology could be used and further stoke excitement. As they set out to test the interest among book publishers, there was no way of knowing they would have a hand in reworking the very economics of the book industry.

*   *   *

Picholine restaurant near Lincoln Center on New York's Upper West Side was known for its whimsy yet traditional entrées. Decorated with lavender walls and plum carpeting, the restaurant offered wild game as the house specialty. The entrées—imported from Scotland—came with a note on the menu that cautioned "birdshot may be present." The cheese station—with regional offerings from the United States, Switzerland, and Spain—got rave reviews in *The New York Times*. And, in 2008, the *Michelin Guide* upgraded Picholine to two stars, marking it as one of the city's best restaurants. That September, too, the chief executives of the six largest book publishers gathered in the restaurant's private room known as the Chef's Wine Cellar. A few months later, many were back together there and again at other restaurants roughly once a quarter for at least a year as they met to discuss their world. Their august literary empire was in trouble. Together, they faced what would become known—once word of their meetings became public—as the $9.99 problem.

A year earlier from that September meeting in Picholine's wine cellar, Amazon had released its e-reader called the Kindle, quickly creating a market for digital copies of books, in part, by using a marketing strategy that offered newly released books and bestsellers for $9.99—substantially lower than what hardcover books were selling for at $30 a copy and roughly what Amazon was buying the digital copies for at wholesale prices from the publishers.

For more than a hundred years, the way things had worked was that a bookstore would buy books in bulk at wholesale prices from the publishers, then turn around and sell them to book customers. The publisher, which did everything from editing authors' manuscripts, to printing and binding the books, to marketing and distributing them, set the "list price," or the price that appears on the cover. The retailer would typically pay the publishers about half of the list price, then turn around and sell it for whatever they liked, presumably more than what they paid.

Then Amazon came with its e-books, which didn't require the same sorts of expenses for the publisher, such as printing and distribution. Instead, publishers sold the digital version to sellers, such as Amazon, for wholesale prices of about $9.99. Instead of marking them up, Amazon was turning them around to customers for the same price or less. Rival e-book sellers were forced to do the same thing to remain competitive in the fast-growing e-book market.

Publishers worried that the fundamentals of their business were going to erode as consumers began to expect lower prices as the norm, deflating the value of the hardcover books that sold for much more. Presumably, the book publishers would eventually be pressured to sell e-books for less at wholesale, squeezing their profit margins along the way. Plus, what was stopping Amazon from cutting out the publishers and working directly with the authors?

The book publishers could see their future, but avoiding it was another thing. Increasingly, some thought they would have to band together to fight back. "We've always known that unless other publishers follow us, there's no chance of success in getting Amazon to change its pricing practices," Carolyn Reidy, chief executive of Simon & Schuster, told her corporate-parent boss, CBS's CEO Les Moonves, in an email in September 2009. She added that "without a critical mass behind us, Amazon won't 'negotiate,' so we need to be more confident of how our fellow publishers will react if we make a move like this." Responded Moonves, "I don't mind the fight as long as we can win it." By December 2009, publishers, such as Simon & Schuster, were delaying the release of their new, hot books for Amazon and the e-book market in order to give the traditional hardcovers a better chance. It was, perhaps, a temporary solution. As the music industry had seen, once the consumer became turned on to a digital version of a medium, pirates were more than willing to step in with a solution.

That was the moment Apple found itself in when Eddy Cue arrived that December in Manhattan—an opportune time.

*   *   *

When Eddy Cue began talks separately with publishers in New York, he expected to do a typical wholesale deal where Apple would buy books in bulk at a set cost, then mark up the price for sale to the consumer. Instead, another structure quickly became apparent, a familiar one to Apple's iTunes and App Store model. In business lingo, this was called an agent model. Apple, as the retailer, would set a revenue share with the publishers. In turn, Apple would then sell the books directly to the consumer at prices set by each publisher. Apple wanted its 30 percent. And it didn't want a delay in timing of titles to favor physical copies. Though the details were still being worked out, John Sargent, CEO of Macmillan, emailed an executive at the book publisher's parent company to share what Apple had just proposed to him. He seemed excited. "Would force Amazon's hand," he wrote.

Publishers pushed for Apple to get a lower cut—much lower. Some suggested 10 percent. That idea wasn't flying with Cue or Steve Jobs. If they couldn't win a bigger slice of the pie, then publishers wanted to ensure they had a bigger pie to share. In exchange for Apple sticking to 30 percent, publishers wanted the ability to price the books higher. When Cue told Jobs about the higher prices, Jobs seemed to agree. "I can live with this, as long as they move Amazon to the agent model too for new releases for the first year," he told Cue in an email draft that Apple lawyers would say he never sent. "If they don't, I'm not sure we can be competitive."

What Apple didn't want was Amazon undercutting it. Unlike many Silicon Valley companies willing to buy growth at the cost of profitability, Apple wanted to begin with a legitimate business on day one, and it couldn't stomach the idea of a race to the bottom on price.

To ensure that didn't happen, Apple's lawyers came up with a way to make sure Amazon couldn't do that—even if publishers couldn't force the bookseller to give up the wholesale model for the agency model. It was a solution that would allow publishers to get their bigger pies while Apple kept its 30 percent.

The fix would give Apple what it dubbed most favored nation status. Or, in other words, book publishers could sell their books for whatever price they wanted, but Apple would get the same price as well. No undercutting. Such an agreement put pressure on the publishers to figure out a way to force Amazon to do what Jobs wanted. In effect, what the publishers would need to do was withhold books from Amazon, unless Amazon went along with the pricing the book companies picked. The Apple bookstore would, in theory, give them that leverage. It would require an industry-wide attack on Amazon.

As Apple raced to get the deals closed by January 27, it was a compelling argument. Still, not a done deal, though. Jobs emailed James Murdoch, an executive at News Corp, the parent company of HarperCollins.

For Murdoch, the economics of its current deal with Amazon were simple enough. For Kindle, the publisher sold wholesale at $13.00 a digital copy. The tech company turned around and sold it at retail for $9.99 to Kindle users. The author got $4.20 if it was a hardcover or $3.30 of an e-book. Under what Apple was proposing, the author would get $2.27 on a book price at $12.99 because less would be left over after Apple's cut. He didn't like the idea of selling books through Apple at $9.99. "Basically—the entire hypothetical benefit of a book without raw materials and distribution costs accrues to Apple, not the publisher or to the creator of the work," Murdoch told Jobs.

Jobs's counterargument was essentially that Amazon had already eroded the book business and Apple was trying to give it power back. He was willing to go higher than $9.99—if the publishers could get Amazon to do the same. It boiled down to three choices, Jobs told him: "1. Thrown in with Apple and see if we can all make a go of this to create a real mainstream ebooks market at $12.99 or $14.99. 2. Keep going with Amazon at $9.99. You will make a bit more money in the short term, but in the medium term Amazon will tell you they will be paying you 70% of $9.99. 3. Hold back your books from Amazon." But, Jobs warned, "without a way for customers to buy your ebooks, they will steal them. This will be the start of piracy and once started there will be no stopping it. Trust me, I've seen this happen with my own eyes."

Rumors were flying that Apple and the book publishers were working on something. Amazon held a meeting with high-profile authors in Manhattan to discuss the idea of cutting publishers out of the business altogether, letting them publish directly through Amazon's digital channels, allowing them to keep 70 percent of the sales.

Five of the six publishers would go along with Apple's proposals, just in time for the Apple event. After the reveal, Jobs told *Wall Street Journal* columnist Walt Mossberg that change was coming to the book industry. When the famed columnist questioned why anyone would buy a book for $14.99 on an Apple device—as was being suggested—when they could get one for $9.99 from Amazon, Jobs told him, "That won't be the case."

"You won't be $14.99, or they won't be $9.99?" Mossberg asked.

"The prices will be the same," Jobs said. "Publishers are actually withholding their books from Amazon because they're not happy."

Some consumers quickly understood what was afoot. "Hello Mr. Jobs," Sethh Humphrey, a student at Wartburg College north of Cedar Falls, Iowa, wrote in an email to the Apple CEO. "I don't really expect a reply from this, but here goes. I am a mac and kindle owner. And with Apple strong arming Amazon into raising e-book prices, this is detrimental to my reading as a college student. You have so much. Wouldn't it be okay for us little guys to have something? If you read this, thanks for your time. Peace."

Jobs *did* read the email. Late that night, he responded curtly, writing, "It's the publishers that are raising prices, not Apple."

"Yes," Humphrey replied the next day, "but the change in prices only comes after your company has let major publishers set there [*sic*] own prices. These publishers must realize that they have almost 100% profit coming in from these e-book [*sic*] because no paper is used. There are other fees and such but still. Greed does not beget most [*sic*], even those at the top."

Not content to let some college student have the final word, Jobs pushed back again. "How do we stop the publishers from setting their own prices and terms?" he asked. "They own the distribution rights to the books, not us. They were already rebelling against Amazon before we ever talked to them."

A random college student in Iowa wasn't the only one to think something was fishy about Apple's dealings.

*   *   *

By 2010, Apple's actions were already attracting attention with the Justice Department's antitrust division, which, along with the Federal Trade Commission, is responsible for enforcing the nation's laws ensuring fair competition. While antitrust cases can be incredibly complex and dense, the basic idea is pretty simple: prevent companies from unreasonably restraining trade. Price-fixing is an obvious no-no, while other actions, such as abusing dominant market position, can be more challenging to prove.

Today's antitrust division at the Justice Department traces its roots back to the days of President Theodore Roosevelt's battles with robber barons. In 1903, his administration created the post of the assistant to the attorney general to enforce antitrust laws. Thirty years later, as the nation's economy grew rapidly, the division was created in order to develop experts in the field as the complexity increased.

The election of Barack Obama as president in 2008 ushered a more aggressive antitrust cop into office named Christine Varney. The Georgetown Law School grad was as polished and experienced as they come and, perhaps most important, unafraid of ruffling feathers. She had served on the Federal Trade Commission during the Clinton administration; then, as a lawyer at the prestigious Hogan & Hartson law firm, she represented Netscape in its monopoly battle against Microsoft in the late 1990s, playing a crucial role in painting Bill Gates as a big bully.

In the summer before her appointment, she had made waves at an antitrust conference in a speech that warned Google was becoming such a behemoth it threatened to stifle innovation, not through the desktop like Microsoft, but through its ecosystem of digital offerings. "For me, Microsoft is so last century," Varney said. "They are not the problem. I think we are going to continually see a problem, potentially, with Google."[1]

Her appointment marked a brief moment for the Justice Department willing to take on high-profile tech companies. Though she would recuse herself from the effort, the department would bring a case against Google *and* Apple.

In 2009, the Justice Department began investigating the hiring practices of several tech companies. The government would come to allege that Apple, Google, the chip maker Intel, Pixar Animation, and the software makers Adobe and Intuit had agreed not to poach each other's workers in an effort to hold down wages.

The companies argued there was nothing improper in such arrangements, saying they needed to come to some sort of agreement to assure each other that they wouldn't try to lure away talent while collaborating on projects together. But in 2010, wanting to avoid a court battle, they agreed to stop the practice as part of settlement agreements.

Then came Apple's digital bookstore.

*   *   *

The drama playing out in the book industry had caught the eyes of Texas's attorney general, Greg Abbott, and Connecticut's attorney general, Richard Blumenthal, who were nosing around into why e-book prices had seemingly increased following Apple's entry into the market. As Apple had gambled, Amazon soon raised its digital prices to match amid pressure from the book publishers. "These agreements among publishers, Amazon and Apple appear to have already resulted in uniform prices for many of the most popular e-books—potentially depriving consumers of competitive prices," Blumenthal said in a statement to the media. "The e-book market is set to explode—with analysts predicting that e-book readers will be among the holiday season's biggest electronic gifts—warranting prompt review of the potential anti-consumer impacts."

As is often the case, the states work closely with the Department of Justice, which can direct more resources to what can become big and costly fights if litigated. When states' initial findings were shared with the DOJ's antitrust office, Christine Varney's chief of staff, Sharis Pozen, was stunned by the apparent collusion among the book publishers. Phone records between executives appeared to show an intense pattern of calls between each other around the time of their secret talks with Apple. Pozen took over in the summer of 2011 as acting assistant AG for antitrust and quickly began moving forward on a case. While in Brussels for an annual meeting between the top-level antitrust officials in the United States and the European Union, she met with the bloc's own investigators, who were also looking into the book industry's moves.[2] Still in the early stages, the two sides had to be careful about what they disclosed before the parties being investigated consented to information sharing. At that point, U.S. officials were mostly focused on the book publishers' actions, but the Europeans hinted they should go back and look at Apple's role in everything. It would become a pivotal moment. As Pozen's team looked closer at emails and records between Apple and the publishers, it became clear to them that Apple played a big part in what they saw as collusion.

On April 11, 2012, Pozen filed their claims against Apple in a federal court in New York City. "Ensuring an open and competitive marketplace allows for innovation, which is good for businesses participating in that marketplace and is good for consumers," she told reporters during a press conference after filing the case. The case would come just months after Steve Jobs died, leaving Eddy Cue and the new CEO, Tim Cook, reeling from the loss of their mentor. They seemed intent on fighting the claims against their late boss.

Three of the five publishers had quickly cut deals with the government while the other two eventually did as well. That left it a yearslong fight between Apple and the government. (Random House was not caught up in the legal battle, because its executives had refused to do a deal with Apple.) All parties denied wrongdoing, with some saying the government's action could give Amazon a leg up in controlling the industry, hurting consumers.[3] Meanwhile, in Europe, Apple chose to settle. Executives seemed less personally affronted with the antitrust officials there than they did with Pozen and her team.

The case was assigned to the U.S. district judge Denise Cote, whose courtroom was located in the Daniel Patrick Moynihan U.S. Courthouse in lower Manhattan. In the days leading up to the bench trial, which she would decide without a jury, she gave the two opposing sides her early thinking, a sort of guidepost that could help lawyers know where they needed to focus their efforts in convincing her otherwise. And her "tentative view" didn't look good for Apple. The judge said after reviewing the filings, she believed the government would be able to prove its case against Apple.

When the three-week trial began, Eddy Cue was the star witness, taking to the stand to make the case that he was working fast to negotiate aboveboard deals with multiple parties. "I felt tremendous pressure to get deals done at this point," Cue told the judge. "Steve was near the end of his life when we were launching the iPad, and he was really proud of it. He was working hard on it. . . . I wanted to be able to get that done in time . . . because it was really important to him."

The government presented evidence revealing collective action by the publishers to strike deals with Apple that they thought would help their fight against Amazon. Telephone records showed a spike in calls between the executives. Internal emails showed them discussing the need to address the threat presented by Amazon.

In its defense, Apple's outside lawyer, Orin Snyder, a high-profile litigator who had represented Bob Dylan and Facebook co-founder Mark Zuckerberg, argued that Apple didn't know the publishers were meeting behind its back and that it was simply trying to enter a new market. "Apple was figuring out how to enter this e-book business at a time when the market was rife with turmoil and not only on the brink of change but, as Amazon's witnesses testified, undergoing fundamental structural change before Apple entered the scene," he told the judge. "Apple was developing, revising and refining post-contract terms, negotiating those deals at arm's length with six and then five separate publishers under a very tight time deadline, and then Apple signed the deals."

It took the judge roughly three weeks to come back with a stunning rebuke of Apple, writing that "without Apple's orchestration" the publishers would not have succeeded in eliminating competition in the e-book pricing. "There is abundant direct and circumstantial evidence, and this Court has found, that Apple knowingly and intentionally participated in and facilitated a horizontal conspiracy to eliminate retail price competition and to raise the retail prices of e-books," she wrote in a 160-page opinion.

Beyond that, the judge singled out Cue's efforts to move quickly to make a deal happen.

"Beyond professional pride, Cue had more personal reasons for making the iBookstore a reality in record-breaking time," the judge wrote. "Cue knew that Jobs was seriously ill and that this would be one of his last opportunities to bring to life one of Jobs' visions and to demonstrate his devotion to the man who had given him the opportunity to help transform American culture."

It was a giant win for the Justice Department, a ruling Apple would appeal but that would hold up.

Both Varney and Pozen would remain thorns in Apple's side for years to come—in different roles as they moved on from the DOJ—as would European antitrust officials who were now convinced the company had a tainted culture. While Apple executives might have still seen themselves as underdogs trying to disrupt industries, they were so much more than that. Their little kingdom was becoming an empire. The iPhone and iPad were gateways to a digital world they controlled, in turn affecting the real world. First music. Then books. Regulators watched that power grow, worrying about a corporate culture that seemed to think it was beyond reproach while at the same time conditioned to control everything.

5 BIG BUSINESS

2010–2011

The same year the iPad debuted, 2010, Epic Games began experimenting publicly with what it might be able to offer on the iPhone. The company had found wild success with its *Gears of War* series and was trying to turn its game-making tool Unreal Engine into a big business, too. In 2008, Mark Rein, Tim Sweeney's longtime confidant, worked to acquire a small video game studio in Salt Lake City called Chair Entertainment that had attracted attention for a game it created for Xbox Live Arcade, an early experiment by Microsoft with a digital download service inspired by Apple's iTunes success. By the very nature of the nascent download capabilities at the time, the games couldn't be as elaborate a production as *Gears of War*, requiring designers to figure out a way to take advantage of limited data size. Chair Entertainment's first offering for the Xbox Live Arcade did just that, a multiplayer underwater battle game called *Undertow*. And the reviews were glowing, especially with what its developers had to work with. "It's vibrant and impressive, making full use of lighting effects to mask the fact that [these] games can't have highly detailed textures due to space constraints," Erik Brudvig wrote in a review for IGN.[1] "Graphically for something so diminutive in file size it's highly stylised, and the levels all have their own unique look and feel," raved Mike Bowden for Eurogamer.[2] In hindsight, it was the kind of studio Epic would naturally look to for trying to figure out how to bring the ambitions of *Gears of War* to the small screen. The idea was to let the founders, brothers named Donald and Geremy Mustard, do their thing and see what they could come up with using Unreal Engine. At the time of the acquisition, they had a game in the works called *Shadow Complex* that was aiming for *Gears of War*–like production value for the low-memory Xbox Live Arcade. The reception was positive, and sales were good, too, for the platform—but nothing compared with full-blown AAA games.

Next, Rein wanted to see what they could do for iPhone. He was looking to better understand the technical end and the economics, too. Games sold in the App Store had a different business model from what Epic was accustomed to, with some games going for just a few dollars and others being offered for free in attempts to make money by selling upgrades or cosmetics inside the game with so-called in-app purchases. After launching the App Store, Apple followed up by extending its payment system to cover in-app purchases, requiring developers to use it, and collecting 30 percent of transactions in the process.

The mandate to Chair was simple: Make a AAA-like game for iPhone. "They said, 'Well, does that mean he wants to make *Shadow Complex*?'" Rein later recalled of their discussion. "We're not putting any restrictions on what kind of game—just come up with an idea and just come up with something that you think would . . . be a really unique and fun game that shows off triple-A production values." They were given six months—much faster than what it might take for an actual AAA game for console. And it had to be much cheaper, $5.99 to download, and also experiment with in-app purchases. That would be a far cry from what Epic was accustomed to. It was normal to spend $20 million to create a console game that might sell for $60. For the Chair team consisting of a dozen people, Epic gave them a budget of about $1.5 million. A couple of weeks later, they came up with a sword battle game that would be called *Infinity Blade*.

In September 2010, Donald Mustard walked onstage at Apple's fall event to reveal the latest iPhone and demonstrated an early version of the game. He showed a 3D world that wasn't normal for a mobile phone as he battled a knight. The three-minute presentation ended with Steve Jobs walking onstage. "It's on a *phone*," Jobs said as Mustard walked off the stage. "That's pretty remarkable."

The game would be released a few months later during the holiday season. Rein's expectations had been modest. "The reason for *Infinity Blade* was a classroom," Rein later told a U.K. gaming conference in the summer of 2011. "We want to learn about this stuff, we want to learn about how the App Store works, we want to learn about micro transactions"—payments made inside the game. They quickly learned a lot, including that the gaming market was changing. *Infinity Blade* was an immediate hit. Epic saw that in terms of return on worker time invested, the game was performing about the same as *Gears of War*, or put another way, compared with working on *Gears of War* for two and a half years and returning "crazy stupid amounts of money," the mobile game was keeping pace, Rein said. Epic received about $11 million in sales in the first six months from the App Store, after Apple's 30 percent cut. And of that money, in-app purchases were making up a hefty share—45 percent of the revenue in the early days. "You'd be surprised how many people pay $50 to buy $50 worth of gold for their $6 game," Rein said. "It's insane." They learned they had to keep investing in the game, spending about $500,000 in the first six months to market and upgrade the offering to keep users interested. "If you want to . . . keep getting bumped in the charts, you have to update them, you have to improve them," Rein said. Still, he wasn't sure that in-app purchases would overtake paid games anytime soon. "We kind of came to the conclusion that for us to make as much money solely on in-app purchase as we've made from the combination of in-app purchase and premium price, we'd have to have thirty times as many downloads as we have now," Rein said. "That's probably not achievable. . . . That's just a ridiculously huge amount to make that kind of money."

Epic would soon see how fast their assumptions would change.

*   *   *

While Epic Games was just discovering what business opportunities awaited through the App Store, another company already knew how important that growing empire was. Whereas Steve Jobs's answer to online music piracy had been selling songs digitally through iTunes, twenty-eight-year-old Daniel Ek was offering another solution through his startup called Spotify: a subscription service that gave users a catalog of songs streamed directly through the internet to a user's computer. Founded in Sweden in 2006, it had quickly become a phenomenon in Europe. In 2010, the company saw its paid subscribers jump from around 320,000 early that year to 500,000 in just a few months. There were many more free users. The success had attracted investors, including Sean Parker, co-founder of Napster, who valued the company at $1 billion. "You get addicted to it," Parker was quoted saying of Spotify. "You end up building a music library that's 100 times bigger than anything you've ever had, and at that point you have no choice—we've got you by the balls. If you want that content on your iPod, you're going to have to pay for it; if you want that content on your iPhone, you're going to have to become a subscriber."[3]

Spotify paid for the music rights by selling ads and paid versions of an account. There were tiers of service: the free, ad-supported one; a $5-a-month version that was ad-free for use on computers; and a $10-a-month account that gave access on an iPhone in Europe.

Not everyone liked the business model. Record labels worried that the buffet of music might cannibalize their sales from, say, iTunes. And some artists were unhappy with their royalties. To enter the U.S. market, Ek had been trying to negotiate with record labels to allow that to occur—deals that would eventually equate to paying out about 70 cents of every dollar earned from streaming to music labels and rights holders.[4] For months, buzz had been building about when Spotify might finally arrive. Ek had been contributing to that hype himself. In early 2010, he announced infrastructure for the United States. "We're in the final stages of setting up," Ek told *Billboard* in an interview. "Yesterday we signed a data center contract, which is huge for us. . . . So, we're gearing up for a U.S. launch. I can't say if it's in one month's time or two month's [sic] time, but it's looking pretty good."[5] He later gave a keynote speech at the South by Southwest Interactive conference. By the end of the year, nothing had occurred. *Wired* ran a profile of the company with a headline that read, SPOTIFY IS THE COOLEST MUSIC SERVICE YOU CAN'T USE.

By the summer of 2011, however, things came together, and Spotify was open for business in the United States. "If Apple's iTunes ushered in digital music's first phase as a large-scale business, then Spotify and other services like it could be its future," Ben Sisario wrote in *The New York Times* about its U.S. debut. Early sales suggested there was a lot of interest in the new way of consuming music. Revenue would skyrocket in the year following its U.S. debut, doubling in 2012 to more than $500 million.[6] In many ways, Spotify represented a threat to Jobs's vision for iTunes, and Spotify had suspected Apple was behind the scenes with its record industry relationships, making it harder for the company to reach the U.S. market.

Parker would suggest publicly that the record labels were talking about how Apple was trying to keep Spotify out of the United States. "It's a very small industry," Parker said at a *Wall Street Journal* tech conference in 2012. "But one of our core competencies is our licensing framework. We are always in negotiation. We're in constant renegotiation. In that process, you hear things. There's a sense that Apple was threatened by what we were doing."

The *Journal* columnist Walt Mossberg asked why Apple couldn't just launch its own competing streaming service. "They probably can," said Ek, who was onstage with Parker. "But the value of Spotify is that we have 700 million playlists."

*   *   *

Phil Schiller, the so-called App Store chairman, was watching TV one night a few days before the Thanksgiving holiday in 2010 and grew unhappy with an Amazon commercial he saw. He dashed off an email to his boss, Steve Jobs, describing the ad for Amazon's digital reader, the Kindle. "It starts with a woman using an iPhone and buying and reading books with the Kindle app," Schiller wrote. "The woman then switches to an Android phone and still can read all her books. While the primary message is that there are Kindle apps on lots of mobile devices, the secondary message that can't be missed is that it is easy to switch from iPhone to Android. . . . Not fun to watch."

Schiller was frustrated. Apple had initially given Amazon an exception from the rule requiring it to use Apple's in-app purchase system because Schiller thought users would be buying books on their Kindle then later access them on their iPhones. "A lot has changed since then," Schiller told Jobs. "We have sold many more iPhones and iPod touch then they have Kindle devices, we have the iPad now as a reading device as well, and their marketing has changed to reflect that more often Kindle app users are purchasing digital books right on their phones."

With its TV ad, Schiller felt, Amazon was making it clear that it was violating Apple's own guidelines. "We should ask them to come back to us with a plan on how they will get their app in compliance with the rules," Schiller added. "Based on our past discussions I expect they will may [sic] choose not to do that. We would then likely have to decide whether to pull the Kindle app from the store or continue to allow an exception to our terms and guidelines for the Kindle app."

Jobs agreed. "It's time for them to decide to use our payment mechanism or bow out," he replied. "And I think it's time to begin applying this uniformly except for existing subscriptions (but applying it for new ones)."

Whether he realized it in the moment or not, Schiller was heading down a path that would see Apple tighten the rules of its burgeoning digital kingdom. The exchange that evening was just one of many that showed how the company was trying to navigate the growing App Economy emerging from the App Store. The iPhone was becoming central to an increasing number of users' media consumption and Apple clearly wanted to capture a piece of that.

Executives began discussing how to offer within the App Store subscription services, and, in turn, extending their requirements for in-app purchases further.

On a Sunday morning in early February 2011, Jobs discussed the idea with Schiller and Eddy Cue, the Apple executive who negotiated the company's troublesome book deals, via email ahead of a meeting. Cue had reached out to nail down what the rules for subscriptions should be going forward. "The basic premise is if the publisher brings a subscriber we get nothing and if we bring the subscriber we get 70/30," Cue told the men. "Apps must offer our in-app subscription offer and can not link out to any other. Publisher can offer subscription through any of their mechanisms (e.g. print, web site)." A key rule, he would add: "The app must have a subscription offer using Apple's recurring subscription (can not link out from the app to any subscription offering)." In other words, developers could not steer their customers to their websites where Apple couldn't collect a 30 percent cut of sales. At this moment, Cue was talking about just newspapers and magazines. In a short time, the subscription idea would encompass digital media—streaming music and video. The anti-steering provision would be at the heart of many battles to come.

Schiller warned them about some of the inevitable blowback, especially Apple's rules that kept developers from receiving typical user information that they might want to use for their own marketing purposes or dealing with their customers directly. "Magazines and newspapers argued that they don't like our offering because they want to get a lot of customer data (mostly name, email, address, phone number) but one of the big things they get by offering an app is a ton of customer data that they never had before—they can learn what stories customers read, how often and long they read, where they are (weather), what teams they follow (sports), when they where born (horoscopes), what companies they follow (stocks and business), what games they play (crosswords, sudoku, etc), what ads they click on, etc. It is very impressive all they can learn about customers without forcing the customer to provide anything that they do not want to share."

Cue raised another hot topic: Should they require in-app purchase for book apps, too? "At first this doesn't seem that bad not to require but the more I think about it, it will be very problematic," he told them. If Apple didn't require book sellers to use the App Store payment system, other developers offering content would naturally want to op-out as well. "What about Netflix, WSJ, MLB, Pandora, etc?" he wrote. "They will all do it. . . . Where do you draw the line? And many other [sic] would want it (e.g. magazines and games). The problem is many can afford 30% but others will say they can't." Even though the

App Economy was just in its early days, Cue already could predict how important their decision was going to be. "This is going to be a huge decision for us," Cue said. "We don't want to lose the apps from iOS and at the same time we don't want to compromise the app experience that we have (e.g. don't have to enter your info or payment everywhere)."

On the issue of books, Jobs would be clear. "I think this [is] all pretty simple—iBooks is going to be the only bookstore on iOS devices," Jobs told Schiller and Cue in early February 2011. "We need to hold our heads high. One can read books bought elsewhere, just not buy/rent/subscribe from iOS without paying us, which we acknowledge is prohibitive for many things."

When they finally hashed things out, the rule changes were sweeping. On February 15, Apple told media companies they would have to offer their digital subscriptions within the App Store app as an option, and, in turn, use their in-app payment system. They could no longer provide links in their apps to their websites to allow purchasing content outside of the app. Essentially, if you were selling digital content that could be used in Apple's app world, then it also had to be sold through the App Store.

The reaction was immediately seen by some as Apple overreaching. News coverage set the stage for companies, such as Amazon, to get kicked off the iPhone if they didn't go along with sharing revenue with Apple. "I think it's going to raise some very serious anti-trust concerns that will be closely scrutinized by the FTC," David Balto, the former policy director of the Bureau of Competition at the Federal Trade Commission, told *The Wall Street Journal*. The tech community was fuming about another change to the App Store guidelines after waiting so long for clarity. "I don't like the precedents that Apple continues to set," Jason Kincaid, a tech pundit, wrote for industry-watcher *TechCrunch*. "The App Store has existed for less than three years, and Apple has been drastically changing the rules on the fly, ruining some businesses and hampering others."

By June, amid scrutiny from the U.S. antitrust enforcers, Apple backtracked—somewhat.[7] Less than a month before the rule changes were set to kick in, Apple quietly told developers it was dropping the requirement that they offer their content for purchase within the app.

Apple's new rules would still prohibit steering customers to external websites to avoid using Apple's in-app payment system. In practical terms, users could go to a developer's website and purchase a subscription and use their iPhones to consume it without Apple getting a cut of the sale. If the developer wanted to sign up new paying customers via its app, it would have to use Apple's payment system and share the revenue. Ultimately, Amazon decided to stop selling digital books on its Kindle app on the iPhone. Users could continue to read their digital purchases made on Amazon's website but not make new purchases through the App Store.

Apple had extended its Walled Garden—maybe not as much as it wanted, nevertheless it had strengthened its anti-steering provision. Still, Schiller had wondered aloud if Apple would really be able to keep charging a 30 percent commission forever. He expressed his doubts to Cue and Jobs in an email that summer. While he called himself a "staunch supporter" of the 70/30 split, in part because it kept things simple and consistent, Schiller told them he suspected new technology—such as web apps—would create increased competition, eroding Apple's ability to get such a large cut. "If someday down the road we will be changing 70/30, then I think the question moves from 'if' to 'when' and 'how,'" he told them. "I'm not suggesting we do anything differently today, only that whenever we make a change we do it from a position of strength rather than weakness. That we use any such change to our advantage if possible." That change could happen, he suggested, once Apple reached making more than $1 billion in profit from the App Store. "I know that is controversial, I just tee it up as another way to look at the size of the business, what we want to achieve, and how we stay competitive," he wrote.

*   *   *

Facebook famously was founded in the Harvard University dorm room of Mark Zuckerberg. The social media platform connected students by desktop and laptop computers. As the company prepared to go public in early 2012, however, there was a problem. The world was consuming the internet in a much different way. Thanks in large part to the success of the iPhone, mobile computing was taking over. In a worrisome sign, Facebook lawyers amended the paperwork for its initial public offering that May, revealing that more than half of its almost one billion users were now accessing its platform through mobile devices. The rapid shift was occurring while Zuckerberg and his team struggled to build space for ads on the mobile version of Facebook—a bad situation for fueling ad growth. "We believe this increased usage of Facebook on mobile devices has contributed to the recent trend of our daily active users increasing more rapidly than the increase in the number of ads delivered," the company said.[8] By the fall, 60 percent of its user base was mobile. The shift had the company racing to adjust, including by acquiring the popular photo-sharing app Instagram for $1 billion. Instagram, along with Spotify, represented a growing number of apps that were blossoming from the App Economy.

By 2012, the App Economy had created roughly 466,000 jobs in the United States and generated, perhaps, $20 billion annually, according to a report issued that year by TechNet, an advocacy group begun by industry luminaries, including John Doerr, the venture capitalist who stood onstage in 2008 with Steve Jobs for the App Store announcement. Apple's control of that economy would prove troubling to Zuckerberg. Quietly, he directed his executives to work to develop their own smartphone as a rival to the iPhone. An executive named Chamath Palihapitiya, who would later become a high-profile venture capitalist, was tasked with leading a team, which he relocated to an unmarked building that required badges different from Facebook ones. The device, code-named GFK after the Wu-Tang Clan member Ghostface Killah, was designed by Yves Béhar, and included a groove for a user's thumb to scroll.[9] Ultimately, the plan was scrapped. Instead, Facebook partnered with the phone maker HTC in 2013—an effort that flopped. The episode underscored the challenges of rivaling Apple in the phone market. To play in mobile, the options were increasingly one of two empires: Apple or Android.

Zuckerberg seemed to acknowledge as much during a presentation with investors that year, noting that he was often asked if Facebook would do its own phone. But if Facebook sold ten million phones, that would be just 1 percent of the market, he responded. "We have a billion people using our product, and we need to make Facebook really good across all of the devices that they use," he said. As much as Zuckerberg wanted to focus on the users in his digital kingdom, the reality was that many were citizens of a higher power, Apple's empire. In the mobile era, for Zuckerberg—just as it was for Daniel Ek and Tim Sweeney—Apple increasingly offered access to an entire digital world of customers. Facebook, Spotify, Epic Games, these companies were building their own worlds—major businesses—within the iPhone empire. It made them citizens of the Apple realm whether they chose to realize it or not. And Apple would do its best to keep them in line.

PART II

WALLED GARDEN

6
AN IRISH PROBLEM

2013

After Steve Jobs died, Tim Cook left his mentor's office at Apple headquarters as it was, something akin to a shrine to visit in those moments when he needed to feel close to his late friend. He famously shared Jobs's advice to him on how to oversee Apple in his absence: "Just do what's right." The two men, by temperament and style, could not have been more different. Jobs provided the vision, while Cook was the operator responsible for the complex, low-cost supply chain in China capable of churning out the iPhone. For years to come, Cook would be harshly compared with the once-in-a-generation innovator. Every public challenge, fairly or not, became a question about whether Cook was failing to live up to the legacy, fretting that Apple was on the decline. Jobs had run Apple mostly as he pleased—a luxury afforded to him after having saved it from near bankruptcy years earlier when he returned as CEO. One of his bugaboos was how to handle Apple's cash from profits—the pile of which was growing exponentially by the day as the company hawked iPhones around the world. Unlike other companies pressured to return cash to investors in the form of dividends, Jobs rejected such things. He seemed to take some sort of comfort in hoarding Apple's profits in case of a rainy day, arguing they should be used for improving Apple. After his death, though, the cash pile had grown to almost $100 billion, seemingly more than the company could spend. Investors hoped—or more like expected—that Cook would take a different approach to returning shareholder value, either proactively or under pressure from Wall Street. In March 2012, Cook held a special call for investors to announce Apple's cash plans. For the first time since 1995, a year before Jobs returned to Apple, the company would begin paying a dividend. In addition, the company would buy back some of its stock, a move that investors generally like because it tends to result in a bump in stock price. In total, the two moves were expected to cost $45 billion. Apple executives wanted to keep some cash on hand in case it was needed. Then there was the money held outside the United States—a complicated knot of a problem that would stalk Cook for years to come. In the previous fiscal year alone, Apple's cash had increased by $31 billion. Of that amount, 77 percent occurred in foreign countries and was being held there to avoid U.S. taxes. As of the end of December 2011, Apple had outside the United States about $64 billion in cash and growing. "Repatriating the cash from offshore would result in significant tax consequences under current U.S. law," Apple's chief financial officer, Peter Oppenheimer, told investors. "We think that the current tax laws provide a considerable economic disincentive to U.S. companies that might otherwise repatriate the substantial amount of foreign cash that they have."

With that statement, Apple quickly became part of a broader debate in Washington, D.C., about creating a so-called tax holiday to allow companies to repatriate their foreign-held earnings under preferential tax rates similar to what Congress had passed in 2004. But as Apple continued to rake in profits from the iPhone, its cash pile just kept growing and growing, a quarterly reminder of the situation— frustrating both investors who wanted the money and U.S. lawmakers questioning why the company was allowed to avoid paying taxes on those gains. By the end of fiscal 2012, the pile of money held overseas had grown to almost $83 billion. By the end of March 2013, cash held outside the United States was $100 billion, most of which wasn't being taxed by any government. Applying Apple's public disclosures, outside groups, such as Citizens for Tax Justice, estimated that Apple would have to pay $35 billion if that money was subjected to U.S. taxes.[1]

Then, that spring, Apple drew more ire when it borrowed billions of dollars to pay for more share buybacks rather than tap its overseas cash. The move allowed Apple to avoid paying taxes on the money. Members of the U.S. Senate had already been looking into companies' tax practices. A *New York Times* story that detailed how Apple was avoiding paying taxes had drawn rebuke from influential lawmakers, such as Senator Tom Coburn, a Republican from Oklahoma. The article said Apple had used legal accounting techniques to route profits through subsidiaries, such as in Ireland, to sidestep income taxes. It noted that in fiscal 2011 Apple paid $3.3 billion in taxes around the world on $34.2 billion of profit, a tax rate of 9.8 percent, while Walmart, for comparison, paid a tax rate of 24 percent, or $5.9 billion on $24.4 billion in profit.[2] "Absolutely, I'm livid about that," Coburn said on MSNBC's *Morning Joe* program. "First of all, we have a tax code. Why should Apple pay at 10 percent and some other company that can't export their technology . . . why are they paying 35 percent?" he said. "What's happening to us right now on our system is everybody that's really successful worldwide is keeping their capital out of here and that capital is being invested somewhere other than America."[3]

For its part, Apple defended its action as legal, noting that it pays a lot of taxes, calling itself one of the top payers of U.S. income tax. It was a defense that Cook was going to take to Washington, where the Senate was holding hearings on the matter—another deviation from the Jobs era. Apple under Jobs showed little interest in politics. Jobs had a saying that if a company required Washington to get into a market, it wasn't really innovating. Under Cook, Apple was spending more on lobbying, still peanuts compared with the likes of Google, and making the effort to develop relationships with key lawmakers. Ahead of the hearing, Cook gave a rare interview to *Politico*, telling the Capitol Hill bible that Apple was doing right. "I can tell you unequivocally, Apple does not funnel its domestic profits overseas," Cook said. "We don't do that. We pay taxes on all the products we sell in the U.S., and we pay every dollar that we owe. And so I'd like to be really clear on that."[4] On the Hill, in closed-door meetings, Cook was greeted enthusiastically by lawmakers who were almost starstruck over the iPhone, according to people familiar with those interactions.

Prior to Cook's testifying before the committee, senators released their extensive findings into Apple's behavior, which would be stressed throughout the day as legal and an example of how tax loopholes should be closed. In great detail, the committee laid out the complex ways in which Apple had used offshore corporations in Ireland to avoid paying billions of dollars in taxes. "Apple is exploiting an absurdity, one which we have not seen other companies use," Senator Carl Levin said. "The absurdity need not continue." He added, "We are trying to shine a spotlight on the practices of a big company." Senator John McCain noted that Apple executives "enjoy reminding the public that the company is likely the largest corporate taxpayer in the United States. However, these same executives fail to mention another less attractive fact: Apple is also one of the biggest tax avoiders in America."

When it came time for McCain to grill Cook directly, the Republican asked if Apple's ability to reduce its tax burden by employing such tactics gave the large company an unfair advantage over smaller U.S. companies without the same muscle. What followed was a master class in obfuscation, of talking around each other, of Cook saying what Apple was doing was fine and to suggest otherwise was an incorrect view. "No, sir," Cook responded. "It is not the way that I see it, and I would like to describe that. The way that I look at this is Apple pays 30.5 percent of its profits in taxes in the United States, and I do not know exactly where this stacks up relative to other companies, but I would guess it is extremely high on the list. . . . We do have a low tax rate outside the United States, but this tax rate is for products that we sell outside the United States, not within."

In essence, Cook was sticking to his defense. Apple was paying the U.S. taxes it was required to pay. McCain gently pushed back, asking why one of Apple's Irish subsidiaries existed. The two men continued to talk in circles for a bit as Cook denied dodging taxes. "When you look at that avoidance or relief of a 35 percent tax burden, which I am sure that we are in agreement is way too high and now the highest in the world, I understand, but you said the purpose of the [subsidiary] is to ease administrative burdens," McCain responded. "But are there certain U.S. tax burdens—isn't it obvious that you are not bearing the same tax burden as if you were bearing in the United States, which then gives you some advantage over corporations and companies which are smaller, which are strictly located in the United States of America? I am not saying that is wrongdoing. But I think you would agree that it gives you a significant advantage."

In short, Cook disagreed. "Again, sir, I have tremendous respect for you," Cook replied. "I see this differently than you do, I believe. What I see is Apple is earning these profits outside the United States. By law and regulation, they are not taxable in the United States. We have set up a holding company to collect these after-tax profits from our different foreign subsidiaries into [the Irish subsidiary]."

McCain replied, "Can you understand there is a perception of unfair advantage here, Mr. Cook?"

"Sir," Cook continued, "I see this as a very complex topic that—I am glad that we are having the discussion, but, honestly speaking, I do not see it as being unfair. I am not an unfair person. That is not who we are as a company or who I am as an individual. And so I would not preside over that, honestly. I do not see it that way."

Cook was not going to give any ground. And as time ran out for McCain, the senator worked to end on a light note. "I am out of time," McCain told Cook. "What I really wanted to ask is why the hell I have to keep updating the apps on my iPhone all the time."

Laughter erupted.

"And why do you not fix that," McCain concluded.

"Sir," Cook responded, "we are trying to make them better all the time."

While McCain and Levin seemed intent on exposing the tax scheme, others appeared more interested in meeting Cook. Senator Rand Paul, a Republican from Kentucky, told the committee they should be apologizing to Apple. "I am offended by a $4 trillion government bullying, berating, and badgering one of America's greatest success stories," he told his colleagues. "Tell me one of these politicians up here who does not minimize their taxes. Tell me a chief financial officer that you would hire if he did not try to minimize your taxes legally. Tell me what Apple has done that is illegal."

When it all ended, it was not the crucifixion some in Washington had expected. Rather, it appeared Cook had used his southern charms to defuse the situation. *Time* magazine's Alex Rogers called Cook's reception "a lovefest."[5] *The New York Times*'s coverage was similar. "Timothy Cook came to the lion's den on Capitol Hill on Tuesday, prepared to face down lawmakers furious over evidence that Apple, the famous company he runs, had avoided paying billions in taxes," wrote Nelson D. Schwartz and Brian X. Chen. "By the time Mr. Cook walked out, the big cats on a Senate committee were practically eating out of his hand."[6] Jon Stewart on *The Daily Show* skewered the Senate. "Apparently there is nothing Apple can do to get us mad at them," he joked. "We could find out they're using kitten hearts to power iPhones and we'd be like, 'Well, if it doubles battery life, I'll take two!'"

There was one place where the Senate's revelations were being taken seriously: Europe.

*   *   *

In a high-rise building in a shabby part of Brussels, roughly a twenty-minute walk from the European Union's executive branch headquarters, sat the offices of its antitrust enforcers. Formally called the Directorate-General for Competition, the group was known among the city's bureaucrats as DG Comp; men and women who work there are modern trustbusters, charged with ensuring fair competition across Europe. In 2013, the European Commission—the executive branch of the EU—was increasingly concerned about member states becoming tax havens for global companies doing business in Europe. Multinational companies were often negotiating tax deals to avoid getting taxed twice—once by the EU country where the sales occurred and again by the nation where the company's European headquarters was located. The problem, according to the commission, was that some companies were getting away with paying no taxes.[7] Companies, such as Apple, had established their regional HQs in Ireland, which didn't prevent companies from shifting profits to tax havens such as the Netherlands, called by Oxfam the "undisputed European champion in facilitating corporate tax avoidance." As the commission debated how to unify the bloc around tax policy, a small group of enforcers—dubbed the Maxforce—was assigned to sniff out suspected tax dodges around the EU. Tim

Cook might have charmed himself out of a jam before Congress, but his appearance there immediately caught Maxforce's attention.

Maxforce was named after the group's leader, Max Lienemeyer, a German lawyer who had cut his teeth during the European debt crisis. A decade-long veteran of the commission, he had been handed a politically fraught task—made even more challenging as the leadership of the commission was turning over. He had hired a relatively young team, many just out of university, to dive into the world of intercompany transactions. One of his first hires was thirty-year-old Helena Malikova, who worked at Credit Suisse Group in Zurich. Early that June, just weeks after Cook's appearance before the U.S. Senate, the commission sent a letter to Irish authorities demanding information about tax favors it had granted Apple. In particular, Maxforce wanted to better understand how Apple's international operations based in Ireland recorded huge profits but apparently paid so little in taxes. What investigators would eventually find was similar to the Senate's conclusions, sparking a massive legal battle that would stretch on for years to come.

Apple had incorporated subsidiaries in Ireland. One was called Apple Sales International and the other Apple Operations Europe to hold the legal rights to use the company's intellectual property to sell and manufacture products outside North and South America. As part of the complicated scheme, investigators found that the subsidiaries were required to make yearly payments to the U.S. office to pay for their share of research and development. In 2011, the payments equaled about $2 billion, mostly paid by the Irish sales arm, equaling much of Apple's R&D efforts that year, according to investigators. Paying a hefty bill like that is beneficial because it allowed the subsidiaries to deduct the expense from its European profits.

The sales arm was set up in a way that it was responsible for selling Apple products throughout Europe as well as the Middle East, Africa, and India. All of those sales were then recorded in Ireland, falling under that nation's tax authority. Apple's tax pros had negotiated deals—first in 1991, then in 2007—with Ireland that meant only the sales of products actually occurring in Ireland would be counted for its tax bill. The rest of those sales, occurring all over Europe, wouldn't. Those sales were being attributed to the "head office"—a misnomer because it wasn't an actual office or place. "There was nothing," Malikova would later recall. "We didn't think it was possible."[8]

As the Senate investigation had shown, the sales subsidiary recorded a profit of $22 billion in 2011, but only €50 million was considered taxable in Ireland. As Maxforce did the math, that meant Apple's sales arm paid less than €10 million for corporate tax in Ireland that year, giving it an effective tax rate of about 0.05 percent. "We never expected to find what we found," Malikova said.

The Senate had given them a road map. Now it was up to Maxforce to figure out how to make a case out of it—especially with Apple *and* Irish officials arguing that they had done nothing wrong. It was going to take political will to soldier through. And a new boss. The head of DG Comp and Maxforce's boss, Joaquín Almunia, was nearing the end of his five-year term.

*   *   *

Margrethe Vestager would become one of the rare European politicians known by name in the United States—even if few could actually *pronounce* her name properly. The September 2014 announcement of her nomination to become the commission's antitrust enforcer came as the EU was grappling with how it wanted to use its sweeping powers that included the ability to fine companies and stop mergers without a court order. Her predecessor had developed a reputation for preferring to correct perceived problems by negotiating with companies, a tactic that could lead to some frustrating outcomes, such as the tech giant Google wiggling to avoid a settlement over its search and advertising businesses.[9] One of the first issues she would need to resolve was what to do with the Apple case. Her bland comments to *The New York Times* sounded like those of any incoming Eurocrat and not someone on the precipice of overturning the Apple cart. "I think that fairness is important in all dealings and of course you have to listen to what people have to say—that goes for companies who feel that other people are misusing their market position and of course for the companies that are being questioned," she said.

Vestager, the forty-six-year-old mother of three teenage daughters, cut her political teeth in Denmark, where she grew up in a small city called Ølgod, the daughter of two Lutheran rectors whose early lessons of right and wrong would shape her worldview. As a young girl, she urged her father to join an invitation-only social club so she could attend an event with a boy she had a crush on, only to be crestfallen when he refused to participate in any group that wasn't open to everyone. "I thought it was extremely unfair because I had thought of this plan, and I thought it would work. And he wouldn't do it for this very abstract reason," Vestager told a reporter. "Since it was painful for me, it also stuck with me."[10] Soon, it would become clear that her views on the role of antitrust rules would sound almost biblical in nature. And she was more than willing to be David in a fight against Goliaths.

As a young adult, she followed in the footsteps of her maternal great-great-grandfather, who founded the Radikale Venstre, joining the Radical Left party and working her way up to party leader in Parliament, then Denmark's deputy prime minister and minister for economic and interior affairs. It was at a time when the economy was reeling, creating tough budget decisions. She publicly supported cutting unemployment benefits and at a press conference defended the action, saying, "That's the way it is." The callousness of her comments seemed like a political misstep that her enemies tried to bludgeon her with. Instead of moving away from it, she gave a speech in which she used the phrase nine times.[11] A labor group gave her a statue of a hand with the middle figure extended, a tribute she would call her "Fuck Finger" and bring to her office in Brussels. She helped inspire the prime minister role on the hit Danish TV show *Borgen*, a political drama about a powerful female politician navigating work and family. Before her appointment to the commission, she rose to prominence in Brussels as president of the Economic and Financial Affairs Council, which was helping steer the bloc through the 2012 euro crisis by creating a continent-wide banking union, among other changes.

Once settled into her new office at the commission's headquarters in the Berlaymont, an X-shaped building in Brussels, she quickly agreed the Apple case was one worth fighting. Her team told her it would be an easy one to build, requiring perhaps a year. They were also pursuing cases against Starbucks and Fiat. It would be surprising to find a company caught in the sights of the DG Comp to take the attention lightly; advocating their position is their corporate duty. Still, there is advocacy and there is war. Silicon Valley companies were coming increasingly under Brussels's scrutiny, giving DG Comp an up-close look at each company's peculiar culture. To the European perspective, Facebook was in your face, super aggressive. Amazon framed everything about being for the consumer. Google always wanted to deal. And then there was Apple, whose executives exuded self-righteousness and an inner belief that they, themselves, were making the world a better place. If that was the case, then it was all but impossible to convince Apple executives that anything they had done might be out of line or unfair.

By the second half of 2015, Vestager had yet to bring a case against Apple, leaving some to wonder if the effort had faltered. The probes against Starbucks and Fiat had ended with piddly results.* The commission ordered the Dutch and Luxembourgian governments to recover €20 million to €30 million from the companies.[12] An almost embarrassingly small amount for all of the attention and headaches the cases had caused. Privately, the commission worried a similar approach with Apple would miss its chance to make a statement.[13] Behind closed doors, Vestager's team was trying to figure out a new theory of the case, one that would be riskier but also could result in a much larger punishment for Apple. The differences in the cases would essentially boil down to how Apple shifted sales reporting to a lower-taxing country while Starbucks and Fiat charged themselves nonmarket prices.

*   *   *

Apple's response to the Justice Department's e-book case should have been a warning to Margrethe Vestager about how far the company would go in defending itself. While other companies might have picked a more expeditious route, quickly settling to put such distractions behind them, Apple hadn't. Instead, the tech company's executives

---

*Both companies would appeal. The Starbucks case was overturned while the Fiat case was confirmed.

seemed to pride themselves on the long fight, especially if they felt wronged. In fact, they had built a battalion of lawyers to defend themselves. Apple's general counsel, Bruce Sewell, had been a firefighter before becoming a lawyer. He described himself as a leader who stayed calm in a crisis. His day typically began at 6:30 a.m. looking for whatever emails had arrived in the wee hours of the morning from Tim Cook—"little cookies," as Sewell liked to call them. In Silicon Valley, Cook was legendary for waking up at 4 a.m. every day and sending off missives before heading to the gym. As Apple's top lawyer, Sewell oversaw a legal department that spanned about 900 people and had an annual budget of almost $1 billion—a good portion of which he used for litigation. In just one lawsuit against Samsung, Apple had 350 lawyers—billing collectively 280,000 hours—in a case that involved reviewing about seven to eight million documents.[14]

Unlike lawyers conditioned to help a company avoid or minimize risk, Sewell subscribed to a theory that a general counsel was there to embrace risk. Or, as he described, *sailing close to the wind.* "You want to get to the point where you can use risk as a competitive advantage—that's the point at which law actually becomes a commercial asset to the company," Sewell would tell law school students. "If you can figure out how to get closer to a particular risk but be prepared to manage it if it does go nuclear then . . . that's a real advantage."[15] As any sailor knows, however, sailing close to the wind can be dangerous and take a boat off course altogether. That's what happened for Apple in the e-book deals. By his own account, Sewell had misjudged things in the company's dealings with book publishers. He hadn't realized the publishers were communicating behind Apple's back, he later recounted. Still, according to Sewell, Cook supported him, telling him after the loss to the Justice Department: "Don't let that scare you. I don't want you to stop pushing the envelope, because that's why legal is an important function in the company."

While Apple spent a fortune on a legal army, its government relations efforts were much more muted. If Apple prided itself on staying out of Washington, D.C., the mindset appeared even stronger in Brussels, an opaque place where bureaucracy thrives and the company had few relationships. "Brussels moves slowly, it's complicated, and the officials here have a 20- or 30-year horizon about how they see their careers," *Politico* once quoted a former minister saying about the bloc. "It takes time to understand and for relationships to mature."[16] By many accounts, Apple seems to have reacted slowly to the inquiry. Roughly two and a half years into the probe, Cook traveled in early 2016 to Brussels to meet with Vestager privately to plead Apple's case. The meeting didn't go well. The platitudes that Cook had relied upon in the U.S. Senate fell on deaf ears at the Berlaymont. Cook is said to have lectured her on tax laws in a way that the Europeans saw as trying to intimidate. Word of the meeting quickly spread around Brussels, another sign in the city's eyes of how out of touch Apple was. "Widely known in Brussels as the worst tech meeting to ever occur," a lawyer close to the commission said. "People say it was pretty damn ugly."

Soon, Washington was stepping into the fight for Apple and other U.S. companies caught in the crosshairs. Robert Stack, a senior U.S. Treasury official, was dispatched to Brussels to meet with Vestager's team. And the U.S. Treasury secretary, Jack Lew, sent a letter to Vestager, which quickly became public, calling on the EU to reconsider targeting U.S. companies. "While we recognize that state aid is a longstanding concept, pursuing civil investigations—predominantly against U.S. companies—under this new interpretation creates disturbing international tax policy precedents," Lew wrote. "We respectfully urge you to reconsider this approach."[17]

Generally speaking, the commission wants to avoid transatlantic drama with the United States. The pressure from the Obama administration would only serve to amplify the pressure on Vestager. Still, she pushed back publicly, telling reporters at a law conference that she wasn't swayed by the U.S. lobbying. "It is the same argument as we have heard before," she said. "Just as it is an obvious right for U.S. tax authorities to tax revenues when they are repatriated, it is also for European tax authorities to tax money that is made in the member states."[18] In her own letter, she responded to Lew to say she wouldn't back down, because doing so risked creating a disturbing precedent as the commission aimed to establish "fair tax competition" within the EU. "The Commission has the duty to ensure that these rules are applied in a non-discriminatory manner by excluding preferential treatment in any form that constitutes incompatible state aid," Vestager wrote. "This does not put into question the U.S. taxation system or go against double taxation treaties concluded by EU member states."[19]

Vestager's case was really against Ireland's practices, though Apple would feel the effects of the outcome. In response, Ireland argued that it gave Apple a deal that was theoretically available to any rival that asked for it.[20] Apple's defense amounted to much of what it had said before the Senate: that it paid its fair share of taxes as long-standing precedent. One could almost hear Cook reciting his lines about how Apple does good and hence can't have done wrong—the sort of self-righteous talk that made Brussels bureaucrats' eyes roll in private as they mimicked his southern twang via their European accents. Their outrage remained hot as they kept coming back to the idea that Apple had basically paid nothing in taxes, especially given its effective tax rate in 2011 was about 0.05 percent—far below Ireland's corporate tax rate of 12.5 percent, already one of the lowest in the developed world. As they continued to work, the rate narrowed even further so that in 2014 it was effectively 0.005 percent by investigators' calculations.

Apple disputed the EU's math, saying it ignored taxes paid in the U.S. (or taxes deferred and eventually paid in the U.S. after tax changes in 2017). Amid the public scrutiny, Apple in 2015 stopped its tax scheme in Ireland. By mid-2016, its foreign cash pile had grown to almost $215 billion.

After spending the summer bulletproofing the case, Vestager was ready to announce in late August the commission's decision that Ireland had, in fact, granted unfair tax benefits to Apple. It would serve as a moment to spotlight the decisive nature of the EU just weeks after the U.K. voted to withdraw from the bloc amid questions about its effectiveness. In a press conference held at the commission's headquarters, Vestager laid into Apple. "The European Commission has today adopted a decision that Apple's tax benefits in Ireland are illegal," she said. "Two tax rulings granted by Ireland have artificially reduced Apple's tax burden for over two decades in breach of EU state aid rules." In plain terms, she spelled out the effects of the arrangement, saying it allowed Apple to pay only €50 for €1 million in profit in 2011 and even less in subsequent years. The commission was requiring that

Ireland recoup a decade's worth of back taxes: €13 billion, or about $14.5 billion, plus interest. A sum that surprised many, especially after the small figures leveled at Starbucks and Fiat earlier.

The decision drew derision from Ireland and Cook, who said the ruling had "no basis in fact or in law." The company called the effective tax rate "a completely made-up number." Ireland's Finance Ministry called its tax rates a fundamental matter of sovereignty and said the commission's decision would create uncertainty. Both vowed to appeal. In the following months, Apple's appeal amounted to nothing less than full scorched earth, accusing the commission of going after a basis of a case to produce a large punitive amount. "Apple is not an outlier in any sense that matters to the law. Apple is a convenient target because it generates lots of headlines. It allows the commissioner to become Dane of the Year 2016," Apple's top lawyer, Sewell, told the Reuters reporter Foo Yun Chee, a reference to the title Vestager had received from a Danish newspaper a month earlier. Furthermore, he called Vestager's case an "absurd theory" and took issue with the suggestion that Apple's Irish subsidiaries were improper because they didn't have any employees on the books, noting that Cook—as CEO of Apple—was making decisions that affected those units. "She is arguing that the base on which we should pay taxes in Ireland is essentially all the profits we generate outside the United States . . . in a place that doesn't do any engineering, doesn't generate any intellectual property for us," he continued.[21]

Apple's roller coaster of an appeal would drag on for years. Its fortress balance sheet, built in part by savvy tax strategy that might—or might not—have been legal, gave it almost unlimited funds to fight off attacks. Yet, for the moment, in late 2016, Vestager was one of the few people to have successfully stood up to the iPhone empire. Her team's dealings with Cook and his team had left many suspicious of the Cupertino company's culture—first the e-book case, then its taxes. There had been quiet grumblings about the power of its App Store. The iPhone empire—growing by more than 200 million devices sold that year alone—had grown too powerful in many people's minds. And Vestager's willingness to fight hadn't gone unnoticed by those trying to find the courage to wage their own battles.

7
BATTLE OF THE BANDS

2014

The late Steve Jobs helped remake the music business with iTunes, selling downloadable songs for 99 cents each. More than a decade later, the legacy of his App Store empire was further disrupting the disruption, leaving the music business on the rocks in 2014. That year, global music industry sales fell to the worst levels in memory, crippled in large part by the decline in digital downloads like those through iTunes. The bright spot for the industry was the rise of streaming music—through either paid services or ad-supported ones like what Spotify was advocating.[1] Yet some music labels weren't so sure this was the future they wanted.

Spotify, after all, lost money. And its losses were only growing.

In 2014, the company's losses almost tripled to €162 million, even as its sales soared.[2] Spotify was increasingly trying to convince many in the record business that its offering was the best defense against pirated music. That by making its service better, Spotify was more likely to convert users from the free service to the paid one, where it generated most of its revenue. That spring, Spotify celebrated reaching ten million paying subscribers out of forty million active users. Its biggest markets were now in the United States and the U.K. Spotify was no longer just a quirky Swedish startup but a global disrupter itself. Growth was coming through smartphone users checking out the service through its free offerings. "People are coming in on mobile and then turning into paid subscribers," Daniel Ek, Spotify's chief executive, told *Billboard* magazine. "About 80% of all people who sign up to Spotify, sign up on mobile first."[3] This created increasing tension with Apple. For years, Spotify faced pressure from Apple to adopt its in-app payment system. Doing so would have meant Spotify would have handed over 30 percent of sales directly to Apple, a cost executives thought the struggling startup could barely afford. By 2014, however, they worried their Apple app might get booted from the App Store if they didn't get with the program. That summer, Spotify began using Apple's app payment system and, in turn, increased the price of its paid subscription by 30 percent to $12.99 from $9.99. If it hadn't been clear just a few years earlier, it was now crystal clear that for businesses—like Spotify, Facebook, and others—mobile phones were their distribution methods. Increasingly, though, the question for Ek was what was next for Apple. Would this giant in digital music stand still as its App Economy ate its lunch?

That spring, Tim Cook, still under pressure to show that he was worthy of being the Jobs successor, revealed his hand. Tapping Apple's massive cash pile, he made the company's largest acquisition in history: $3 billion to buy Beats Electronics. The music brand was best known for its stylish headphones, but the real potential lay in its nascent streaming music service and co-founder Jimmy Iovine, a longtime music executive with the charisma and cachet required to add firepower to a new streaming war. "These guys are really unique," Cook said in announcing the deal. "It's like finding the precise grain of sand on the beach. They're rare and very hard to find."

*   *   *

Apple's annual Worldwide Developers Conference was a key way Phil Schiller communicated all of the coming software changes to the iPhone empire so third-party programmers could be ready for the new iPhone that fall. The week of events in San Francisco was a highlight for many aspiring developers and a moment to show off their stuff to Apple. The big draw that June 2015 was the expectations that Apple was going to reveal its answer to Spotify's streaming juggernaut. But first, in the more than two-hour presentation that would touch on all aspects of Apple's ecosystem was a demonstration by Epic Games. Billy Bramer, a game-play programmer, took the stage to talk about how he was using Apple software to prepare an upcoming game from Epic: *Fortnite*. Video of the game showed a cartoon-like world where multiple players used axes and other tools to gather materials to build forts and fight off monsters that came at night—hence *Fortnite*. "If you've ever built a pillow fort or battled imaginary monsters with your friends you already know how to play Fortnite," he told the crowd. "This is the end of the world scenario you've been training for since you were a kid." The game was expected to be out by year's end in a beta, or for testing, he said.*

Later Eddy Cue, the senior Apple executive who had negotiated the book industry deals, took the stage to reveal the fruits of the Beats purchase: Apple Music. But before doing so, he pointed to Schiller in the audience to wish him a happy birthday. The audience cheered as if he were a celebrity while he shook his head in modesty. The news of the day: The streaming service would cost $9.99 a month. Cheaper than Spotify's Apple app or what the Swedish company had been charging before it was pressured to use Apple's payment system.

*   *   *

Soon after Apple Music began in 2015, Spotify launched its own countermeasure—a test really. It sent an email message to iPhone

---

* Not on stage but at the rehearsal was Tim Sweeney. A few weeks later, he would email Tim Cook, urging the Apple CEO to think about separating the App Store curation from compliance review and app distribution. "The App Store has done much good for the industry, but it doesn't seem tenable for Apple to be the sole arbiter of expression and commerce over an app platform approaching a billion users," Sweeney told Cook. In his thinking, Sweeney suggested, compliance could be limited to things such as safety and privacy and compliant apps could be allowed to be distributed without restrictions on "engaging in commerce directly with users." He concluded: "It would be extremely positive for Apple to take this approach proactively before the topic is overly complicated by opposing political, regulatory, moral and competitive forces." Cook looked at the email and forwarded it to Phil Schiller, asking: "Is this the guy that was at one of our rehearsals?"

customers encouraging them to cancel their subscriptions through the App Store and sign up anew via Spotify's website. "In case you didn't know, the normal Premium price is only $9.99, but Apple charges 30 percent on all payments made through iTunes," the email read. "You can get the exact same Spotify for only $9.99/month and it's super simple." What followed was a step-by-step explanation for shutting off auto-renewal through Apple and transferring accounts to the web version—something that could only occur once the subscription lapsed.[4]

Such messaging was seemingly banned within Spotify's Apple app under App Store rules crafted years earlier. What Ek's team was doing was an experiment in the United States and the U.K. to see how dependent they had become on Apple's in-app payment system that automatically renewed subscribers each month. The results weren't good for Spotify. Emails directing users to sign up outside the app weren't a viable alternative to in-app communications, the company concluded. Spotify was stuck between a rock and a hard place.

The experiment also accomplished another ambition: getting Spotify's dilemma greater attention. The email quickly spread through the business media along with the caveat that such messaging by Spotify would probably be banned in its own app by Apple, raising antitrust questions, especially following Apple's e-book case. "I think Apple is getting close to the line, if it hasn't crossed it, in its dealings with Spotify," Tim Wu, a professor at Columbia Law School, told Bloomberg News. "It's a different game when you have market power and you have a direct competitor you're telling can't advertise their prices. That's pretty gnarly."[5]

Fighting Apple's digital empire in a U.S. court on its own seemed like a losing proposition. Since the launch of the iPhone, Apple had faced seemingly endless claims that it was overreaching with its Walled Garden—claims the iPhone maker could drag out with its deep pockets in costly litigation and that seemed destined to fizzle out. One case accusing Apple of using a software update to secure a monopoly over the digital music market with its iPod had dragged on for ten years, ending in a federal jury in Oakland siding with Apple.[6]

Instead of the courts, Ek's team turned to Washington for help, hiring lobbyists and outside legal firepower to begin jawboning law-makers and regulators. It was a fairly standard tactic for a company like Spotify, feeling threatened by a bigger rival, to push a key lawmaker to take interest in its case. Perhaps one might write a letter to the Department of Justice or the Federal Trade Commission, asking about the practice of a rival, giving government lawyers some cover for taking interest against a politically connected giant, such as Apple.

The Spotify lobbying team began making the rounds to quietly talk with lawmakers about how Apple was threatening to stifle competition by using its size to strike unfair deals with record labels for its new streaming service and by collecting the 30 percent cut on all subscriptions. It was one of many steps to plant the seed of an idea that Apple was misusing its market power in anticompetitive ways. They took their case to the House Judiciary Committee chairman, Bob Goodlatte, a Republican, and others in the House as well as members of the Senate antitrust subcommittee.[7]

Spotify also hired an idealistic lawyer named Jonathan Kanter to press regulators. Then in his early forties, Kanter had begun his legal career at the Federal Trade Commission and was part of a group of young lawyers—including Wu at Columbia—inspired by the late Supreme Court justice Louis Brandeis, who came to prominence during the Great Depression amid concerns about the tyranny of monopoly power. Brandeis had warned of the "curse of bigness," where a profitable company uses its money and influence to grow bigger and bigger, quashing competition and further empowering itself along the way. "We can have democracy in this country, or we can have great wealth concentrated in the hands of a few, but we can't have both," he is famously attributed as saying.

But his thinking had gone out of style. Antitrust laws have swung from extremes for almost a hundred years, a tension between populist protections and freewheeling markets. The late 1970s saw a reemergence of a laissez-faire approach, led by Judge Robert Bork, a former Yale University law professor who advocated that government involvement in business only removed efficiencies from the market and harmed the consumer. The new mantra among judges, regulators, and lawmakers became about ensuring the consumer had the lowest price possible.

Essentially, if a company could argue that the consumer wasn't being hurt by higher prices, then its monopoly power wasn't so bad after all. Yes, a little competition was good, but bigger was better from an efficiency standpoint. That view was good for Apple. It was selling songs for 99 cents and offering an App Store where most of the programs were free and easy to use. The consumer was benefiting, Apple argued over and over. And its sales proved their support. But by 2010, the pendulum was returning to Brandeis's view. Kanter and a new group of young legal scholars, worried about the power of Big Tech, were breathing new life into how the nation's antitrust laws could be applied to the digital realm.

As a kid, Kanter had grown up in Queens, New York, a neighborhood full of schoolteachers and taxicab drivers who instilled in him a view he would hold later in life about the American dream: providing opportunities to realize a better life for future generations. In his mind, antitrust law was about making sure opportunities were available for all.[8] He earned a law degree at Washington University in St. Louis. After the FTC, as a partner at a white-shoe law firm called Cadwalader, Wickersham & Taft, Kanter developed a reputation for trying to make the case to regulators and lawmakers in Washington for tech companies against their rivals becoming too big. It could create odd bedfellows. Microsoft, ensnared in a battle for web browser dominance, hired him to help combat Google's efforts in 2007 to acquire the online ad firm DoubleClick. Years later, the deal would be seen as pivotal in allowing Google to evolve into a digital advertising giant. He helped other tech players lobby power brokers as well.

Kanter and Spotify's in-house team made their way to a meeting with the FTC to present their case of why Apple's behavior was a violation of antitrust law. The overall theory was similar to the one the government pursued against Microsoft years earlier: that Apple had an ecosystem that was leveraging dominance in one space to distort competition in another.

But Spotify left disappointed. There was either a lack of understanding of the power of the App Economy or a lack of an appetite to go after Apple—especially after Cook had basked in the glow of the U.S. Senate in the tax matter that seemed like such an open-and-shut case. Apple was seen as an innovator, and the Spotify team kept hearing, Why can't Apple charge money? The Obama administration seemed to be enamored with everything Silicon Valley. Many top former Obama officials, in fact, would end up working for the biggest companies in the months and years ahead.

*   *   *

As Spotify faced uncertainty with Apple, Daniel Ek was also preparing the company to go public, a major milestone for a startup when founders and investors can cash in on years of work. Part of that effort was hiring a general counsel to prepare the company for such an undertaking. He turned to Horacio Gutierrez, who had months earlier become general counsel at Microsoft after spending several years handling the tech company's intellectual property issues. Insiders would later say they didn't recruit Gutierrez for his considerable experience in antitrust law. Still, he would soon guide the company on a yearslong path to fight Apple's hold on its empire and along the way become one of the biggest threats to Tim Cook.

Gutierrez's career arc would prove beneficial in more ways than one. He was leaving Microsoft, where months earlier he had been promoted to the company's general counsel. At most companies that would make him the top lawyer, but he lived in the shadows of his longtime boss, Brad Smith, who had been promoted to president and chief legal officer. The two men had long worked together, going back to the early days of both of their careers. Almost twenty years earlier, Smith had spotted Gutierrez's talent while visiting Microsoft's offices in Fort Lauderdale.

From an early age, Gutierrez stood out. Born in Maracaibo, Venezuela, he was the son of a lawyer. At age sixteen, he began law school and was a partner in a firm in his native home by age twenty-eight. He came to the United States to study at Harvard University as a Fulbright scholar, earning a master of laws degree. After moving to Florida, which required a juris doctorate from an American Bar Association–accredited law school, he earned another law degree from the University of Miami.[9] After that, he soon ended up at Microsoft shortly after the Justice Department brought its anti-trust claims against the company. He played a role in that defense, and Smith asked him to deal with a similar legal battle with the European Union, putting him smack in the middle of the Brussels fight with the U.S. tech giant.[10] The company spent years fighting governments around the world over the claims that it was unfairly competing. It settled the U.S. case in 2001. Appeals and legal battles in Europe would drag on until the end of 2009.

It was a taxing fight for the company. While Microsoft negotiated settlements that left the giant mostly intact, the company seemed to stumble for years to come, missing out on digital music trends made popular by Apple's iTunes, then smartphones and online search. Microsoft's co-founder Bill Gates would later lament that the antitrust battles sapped his attention at a generational crossroads of technological change. "There is no doubt that the antitrust lawsuit was bad for Microsoft," Gates said at a conference. "We would have been more focused on creating the phone operating system, so instead of using Android today you would be using Windows Mobile." He added, "I was just too distracted. I screwed that up because of the distraction." Others inside Microsoft had concluded something similar. Smith would become Microsoft's top lawyer after proposing in 2001 to the board that it settle its battles. During his job interview, he famously placed a slide on the screen that read, "It's time to make peace." He would say later the battles came with opportunity costs. "Every hour you spend doing one is an hour you don't spend doing something else," he said. "That's sort of like the laws of physics somewhere. If you have to spend a lot of time dealing with an antitrust issue, it will mean that there is a lot of time that you are not spending on other things. And you can't possibly know in advance what those other things would be.

"It's only a decade later when you see the parts of technology that maybe passed you by. And you can ask yourself, what if? What if I wasn't spending all that time in a deposition or getting ready for it or dealing with the controversy around this? Would I have recognized this other shift? Would I have done a better job of jumping on it?"[11]

Gutierrez rose up through Microsoft during that era. He excelled for years overseeing the company's intellectual property, fighting the biggest tech companies around the world over complicated patents, and raking in billions for the tech giant. By the time he was announced as Spotify's new general counsel in March 2016, he was a fully tested corporate lawyer. And within weeks, he saw firsthand how tough of an opponent Apple would be.

*   *   *

Horacio Gutierrez wasn't anti-Apple. In fact, he was a fan of their products. While working for Microsoft, he had banned his kids from owning iPhones. The day he decided to leave Microsoft, Gutierrez went out and bought an iPhone and Apple Watch. Still, as he settled into his new job at Spotify, he couldn't help but be surprised at what Apple was able to do to control its empire. He probably could imagine how Microsoft, amid all of its scrutiny, could never have gotten away with such controls. It would have been like Microsoft selling its word processing program Word while limiting access on its Windows platform to rival WordPerfect while also taking a cut of the competitor's revenue as well. Regulators would have been howling in the streets.

Gutierrez was hired to prepare Spotify to go public but very quickly he realized that meant dealing with an existential risk hanging over the company: *Apple*. To go public, Spotify was going to need to convince investors that there was a pathway to profitability. The economics of the streaming business were rough. Basically, most of the revenue generated went toward music licensing deals and the rest went to Apple. Gutierrez was going to need to renegotiate—everything.

It would be tricky.

Late that May, Daniel Ek's engineers sent an updated version of the Spotify app to Apple's App Store that made a dramatic change: New users could no longer upgrade to paying subscriptions inside the app. Spotify was turning off in-app purchases for new users. Instead, it created an "email me" button for users to click to be sent information about an opportunity to upgrade at a discount. In response, Apple rejected Spotify's app update, adding to its rules not only barring links to outside-app payment services but also any "calls to action that direct customers to purchasing mechanisms other than IAP."

In subsequent days, Apple announced ahead of its annual developer conference that it was changing the terms of the in-app purchase system when it came to subscriptions. Instead of taking 30 percent, it would reduce its cut to 15 percent for subscriptions after their first year. The move was announced in a rare interview by Phil Schiller, Apple's marketing chief, with *The Verge* and dubbed "Subscriptions 2.0"—an effort by Apple to get game apps and others to move toward a subscription model. The interview marked Schiller's increased involvement in the day-to-day management of the App Store. He'd always been heavily involved, but now marketing of the store fell under him. "With all of these changes happening to the App Store, it's worth asking: why now? Or rather, why not sooner?" Lauren Goode wrote. "Schiller's newly expanded role and the recent focus on the App Store might easily suggest that it wasn't being paid enough attention before, but Schiller denied that was the case. Instead, he says, the changes are a sign of the times and the success of the store, which he calls 'the best place ever created for distributing software.'"

In the background, Spotify was growing concerned. Its app was on ice. Engineers were accustomed to updating the app every week or so. The sticking point appeared to be their effort to funnel customers to their website, so they kept trying new ways of doing that. Spotify's team tried to look to the App Store rules for guidance on how it could inform its own customers about how to subscribe to Spotify. They felt as if the rules kept changing. Not being able to sign up new users was a problem for Spotify's business model. It hooked new users with the free service with the aim of turning them into paid subscribers down the road. If Spotify couldn't lure fresh fish into the funnel, it was losing out. At the time, more than 90 percent of revenue was coming from the paid tiers. On top of that, there were bugs in the app that weren't getting fixed and brands that wanted to run marketing campaigns that needed to be uploaded.

Gutierrez thought he might be able to negotiate with Apple. He traveled to Cupertino to meet directly with Apple's top lawyer, Bruce Sewell. The two men had known each other for years. But it soon became clear that Phil Schiller and others were calling the shots.

After several weeks, Gutierrez sent a blistering letter to Sewell.

"This latest episode raises serious concerns under both U.S. and EU competition law," Gutierrez wrote. "It continues a troubling pattern of behavior by Apple to exclude and diminish the competitiveness of Spotify on iOS and as a rival to Apple Music, particularly when seen against the backdrop of Apple's previous anticompetitive conduct aimed at Spotify . . . we cannot stand by as Apple uses the App Store approval process as a weapon to harm competitors."[12]

The letter was also distributed to certain congressional staffers. Soon, Senator Elizabeth Warren, a Democrat, was grouping Apple in with big tech behemoths that she said were practicing anticompetitive behavior. In particular, she took aim at Apple for having "placed conditions on its rivals that make it difficult for them to offer competitive streaming services."

"Google, Apple and Amazon have created disruptive technologies that changed the world, and . . . they deserve to be highly profitable and successful," Warren said. "But the opportunity to compete must remain open for new entrants and smaller competitors that want their chance to change the world again."[13]

Apple responded forcefully, distributing a letter it sent to Spotify to reporters, accusing the streaming service of wanting special treatment—a claim it would turn to often as developers spoke out against its rules. "We find it troubling that you are asking for exemptions to the rules we apply to all developers, and are publicly resorting to rumors and half-truths about our service," Sewell wrote. "Spotify's app was again rejected for attempting to circumvent in-app purchase rules, and not, as you claim, because Spotify was simply seeking to communicate with its customers." Spotify responded by posting on Twitter that its app included "no offer, no purchase, no link to anywhere at all." A screenshot simply read, "You discovered a Premium feature! You must have a Premium subscription to unlock it."

After about four months, Apple's stonewalling on approving Spotify's app stopped; the team heard from an Apple business development executive who suggested that if Spotify turned on a feature related to Siri, the stream would finally get approved. That set the Spotify team racing to make that change along with all of the other ones deemed critical after weeks and weeks of backlog. It was a triage situation. They didn't know if, or when, they might get another chance to update, and they needed the app working properly. This could be a one-off shot for a while. Internally, they called the update the "Unicorn." Once they submitted, they waited with bated breath. When word came through, it was a moment of joy. In his Manhattan office, Gutierrez gave an awkward high five to one of his deputies.

It would be only a brief reprieve.

*   *   *

Following her stunning case to make Apple pay $14.5 billion in back taxes to Ireland, Europe's chief antitrust officer, Margrethe Vestager, came to New York City to give a high-profile speech at a TED conference. In the speech, titled "The New Age of Corporate Monopolies," she defended the need for rules that protect competition, warning that those who gain power are often reluctant to give it back. "The problem is that sometimes, for businesses, competition can be inconvenient, because competition means that the race is never over, the game is never won," she said onstage. "No matter how well you were doing in the past, there's always someone who are out there wanting to take your place. So the temptation to avoid competition is powerful. It's rooted in motives as old as Adam and Eve: in greed for yet more money, in fear of losing your position in the market and all the benefits it brings."

Vestager's tough actions with Apple had not gone unnoticed at Spotify. Even before Horacio Gutierrez had been hired as Spotify's chief lawyer in the spring of 2016, its lawyers had been working to see if there might be interest in Brussels in taking up Spotify's fight. In many ways, it made more sense that they might receive a warmer welcome than they did in Washington. Spotify was founded and headquartered in Sweden, a member of the European Union. Daniel Ek was one of the few European tech entrepreneurs to have reached Silicon Valley–like success. "As a victim of a crime, it's easier to go to the authority where you reside," one Spotify executive summed it up.

Vestager's team was more than willing to hear what they had to say, but Spotify team members felt the response was cold and guarded. They weren't going to get a free pass just because they were European.

"What are you complaining about, Spotify is successful?" they would hear.

"Antitrust law should not be there to look at the bones of dinosaurs and say, 'Ah, a meteor killed it or the ice age killed it,'" Harry Clarke, a Spotify lawyer who would become a key deputy to Gutierrez in the antitrust fight, argued. "It should be there to help competition thrive."

Gutierrez's arrival turbocharged the efforts to get the commission looking into Apple. He had worked in Brussels, he knew people there, and, more important to the team, he clearly understood how to get things done in a city where, for outsiders, it can be confusing and opaque. So, while Apple was depending upon Tim Cook's celebrity to pressure EU officials, Ek had essentially hired a ninja. One of the first moves Gutierrez made was hiring an old nemesis from his days at Microsoft, a lawyer who had been on the other side of the fight, antagonizing the tech giant, Thomas Vinje, a partner at the law firm Clifford Chance. During the yearslong battle, Vinje, who grew up in Kirkland, Washington, not far from Redmond, where Microsoft would later be headquartered, was asked by a reporter if fighting the giant was a job or something more. "To some extent it's what the Germans call a *Glaubenssache* [a matter of conviction]. I would have difficulty being on the other side of these issues," he responded. "I'm not averse to representing dominant companies under certain circumstances. I have no innate objection to dominance if it's achieved through innovation and legitimate competitive behavior. But at the same time, I have a deep conviction that there has to be a level playing field. Companies must be able to compete on the merits of their creativity and hard work. I'm getting paid well—I don't deny that. But I feel good about it. I'm an idealist. Who controls the Internet has great social and economic implications."[14]

Across the Atlantic, Gutierrez was happy to find that Jonathan Kanter, the outside lawyer, was already on the job. The two men had worked together before on Microsoft issues. Kanter had experience in the kinds of campaigns Spotify was embarking on—a multifront attack to show the world the Apple that Spotify saw. A monopoly case against Apple would be about more than legal arguments and courtroom appearances. As Microsoft had seen, a successful fight involved communicating and changing public perception. Microsoft's competitors—IBM, Oracle, even Apple—had helped seed regulators with arguments against its business practices, making and building the case for why its conduct was unfair. Now Spotify wanted to do the same thing about Apple.

Little did Spotify know, but Vestager's team had been looking for a case like Spotify's, a face to put on a complaint. It was one thing to suspect Apple was overstepping but another thing to prove it. And it seemed Spotify was more than eager to open up its secret financial records to provide a road map for how the App Economy worked. In conversations with Vestager's team, Spotify began conducting a new round of experiments to show the effect of Apple's anti-steering provision on its ability to convert free users to paying ones. Antitrust cases are often fought on theories that something is harming a rival or consumer and that if not for the offending behavior, things would be better. A judge or jury has to believe the argument. In this case, however, Spotify had a unique way of showing how things might be different if not for Apple's behavior. Spotify operated in a parallel universe of Google's empire, where Android, at the time, didn't prohibit steering users to off-site payment methods. This would allow Gutierrez and Vinje to show what life was like with and without the provisions—in real time. Using the Android platform, Spotify created two versions of its app: a regular Android version and a modified version that had an Apple-like experience that didn't allow in-app users to conduct the upgrade payment through a web-based checkout located on its website. It would then track the rate of upgrades among the two groups within the following months.

The first experiment was conducted in May 2018 on users in Spotify's five largest European markets—France, Germany, Italy, Spain, and the U.K. A second experiment was run in December on a wider group—not just the five largest European markets, but also Australia, Brazil, Mexico, and the United States. This time, users were allocated randomly to four groups with different user experiences for upgrading to premium accounts. The control group got the standard Android experience, which included a "get premium" button that, with a click, directed users to the web-based checkout page for payment. A second group got a variation of the standard Android app except instead of clicking a button to a web-based checkout page within the app, they were sent to a web browser to make payment. The idea was to mimic what an Apple customer might experience if Apple allowed users to be redirected outside the app for payment. The third group experienced the Apple-style app experience—similar to the May experiment. And the final group was given even less information in line with changes to App Store rules at the time, including not receiving any information on the price of premium subscriptions. In both experiments, the results showed fewer people upgrading among the groups not provided information about alternative methods. Using the data, the commission would conservatively estimate that Apple's restrictions were resulting in Spotify losing out on 20 percent of its in-app users upgrading. Or as it would later conclude, millions of "users got lost in the subscription process and did not end up subscribing," and millions more had "an inferior user experience."

SUPER APPS

2017

On 175 acres of land near Interstate 280 in Cupertino, a shiny flying-saucer-like structure took shape. Made of glass and metal, the three-million-square-foot ring was nearing completion, the final vision that Steve Jobs had for Apple before his death: a new head-quarters, dubbed Apple Park, inspired by his walks amid the gardens in London. The four-story building had floor-to-ceiling windows, surrounding a thirty-acre landscaped park filled with apple trees and walking paths. The cold glass walls and warm wood furniture found inside gave the same vibe as an Apple Store in the mall, but bigger—so much bigger. Its completion was coming as Apple prepared to cel-ebrate the tenth anniversary of the iPhone with a new version of the device that fall, a launch Tim Cook was betting would reignite customers' interest.

For Cook and Apple, the year 2017 could be something of a fresh start. It needed to be. They were coming off a very bad 2016 when, for the first time, iPhone sales fell compared with the previous year. Since launching the device, Apple had known nothing but growth, fu-eling record profits and its fortress balance sheet. More important for Cook's job security, the promise of growth had kept investors happy, too. In 2011, Apple knocked Exxon off as the world's most valuable publicly traded company as its stock soared. By 2015, Apple's stock still wasn't just the top valued; its market value had become at least double No. 2 Exxon as it approached $800 billion.[1] Growth that year had been fueled by a new generation of iPhones that had helped Apple move into China in a bigger way. The September 2014 debut, for the first time, saw two form factors of the iPhone, one with a 4.7-inch screen and a larger one with a 5.5-inch display called the iPhone 6 and iPhone 6 Plus, respectively. Physical sales that fiscal year of the iPhone rose almost 40 percent to a record 231 million units. The larger-screen phone was Cook's gambit at appealing to the Chinese market along with a deal he signed in late 2013 with China Mobile, the world's largest mobile carrier with more than 700 million subscribers. That was a customer base seven times the size of Verizon, the largest carrier in the United States, and marked a yearslong effort to improve Apple's position in the important consumer market.[2] Apple had long struggled in China amid low-cost rivals, and the bigger screen helped change that. China iPhone shipments soared 45 percent in 2015.[3] At one point, Chief Financial Officer Luca Maestri was especially effusive after finishing a quarter that saw the major gift-giving holiday of Chinese New Year helping the company sell more iPhones in greater China than in the United States for the first time ever. "The progress we've made in China has been remarkable and we continue to make incredible investments in China," Maestri said. "The growth rate in China is significantly higher than most parts of the world. In the short term, we don't expect China to become bigger than the U.S. but over the long arc of time, you could certainly draw that conclusion."[4]

The gains in China would be short-lived, however. The follow-up offering of the iPhone 6 family in September 2015 was essentially the same devices. Sales that fiscal year dropped 8 percent to 212 million phones—still a stunning number compared with where Apple began just a few years earlier, but a very tangible sign of what critics had been warning about for years: Apple was reaching a saturation of the smartphone market. The hope that China could keep the party going didn't work out as planned; iPhone shipments plunged 21 percent. Its market share fell behind companies that many Americans had probably never heard of: Oppo and Vivo, not to mention bigger

Chinese players like Huawei and Xiaomi.[5] As Apple approached a billion iPhone active users that year, it seemed, there were only so many more new smartphone buyers out there. Essentially, the iPhone business had become a replacement business for customers. And the updates between phones each year had become iterative, meaning users weren't feeling compelled to buy the newest one each fall.

Looking at the company's revenue streams that fiscal year, one bright spot stood out: what Apple called services. Users knew it as the money they were spending on the App Store, iTunes, and other digital offerings in the Apple empire. In fiscal 2016, services had grown to become Apple's second-biggest revenue stream after iPhones, overtaking money made from the sale of computers. Whereas iPhone sales were down, services revenue rose 22 percent. The category was up 50 percent from just three years earlier. As Cook looked at Apple's business, he was investing heavily on bets that future technology would fuel massive growth, putting R&D dollars toward developing virtual-reality-like glasses and electric, driverless cars. But those were gambles that were years away at best, if ever. To satisfy Wall Street's demand for continued growth, Cook saw the near-term future in digital services. He saw a future in the App Store—the biggest source of revenue within services. That's where he would squeeze the organization for growth. "Services are becoming a larger part of our business, and we expect the revenues to be the size of a Fortune 100 company this year," Cook told investors that January. "Our services offerings are now driving over 150 million paid customer subscriptions. This includes our own services and third-party content that we offer on our stores. We feel great about this momentum, and our goal is to double the size of our Services business in the next 4 years."

In many ways, Apple's iPhone business was becoming like an annuity. It was headed toward a world where Cook would sell roughly the same number of iPhones each year to replace those whose phones needed to be replaced. Some years, when Apple could offer a new must-have feature, it might see a bigger burst to sales, but likely at the expense of the following year's results that were pulled ahead. What Cook was signaling was that Apple's growth would come within its digital empire. Apple had built a digital market approaching one billion users, living and breathing on their iPhones every day. The App Store was the gateway to those users.

Behind the scenes, Cook's attention that year would turn even more intensely on services, holding monthly meetings with service team members, demanding they update him on which apps were selling well, and digging into other minutiae of the business.[6] It could help explain why Spotify executives in 2016 felt the tightening of Apple's rules aimed at ensuring compliance with its payment system where it collects a share of all sales. Or why Apple was trying to encourage game makers, such as Epic Games, to roll out recurring subscription models. Or why Apple itself was coming out with digital offerings such as Apple Music.

Apple was still coming to terms with how the App Economy had transformed its business from device seller to service provider. For years, for example, senior executives had debated whether to open some of the original features of the iPhone to Android users in an effort to build out its digital services. In 2013, when a rumor circulated that Google was interested in acquiring a popular messaging app called WhatsApp, Apple executives discussed offering up the iPhone's own message system, dubbed iMessage. The text-based messaging system, which used the internet to communicate, had become best known for presenting in blue bubbles among iPhone users—a cue picked by engineers as they developed it internally to let themselves know easily when working with iMessages rather than other texting formats. It stuck when launched to the public and quickly became a status indicator. Non-iPhone users—sending messages via traditional SMS and MMS—would turn the conversation green. App Store iMessage games would evolve as would additional features such as video and photo sharing. "I had no idea that there would be a cachet or like, 'Ugh green bubble conversations,'" Justin Santamaria, an Apple engineer who worked on the original feature, said years later. The social pressure was especially apparent in younger users. Apple iPhone's share of users aged eighteen to twenty-four grew to 44 percent in 2017, much higher than those older, and would only climb higher in years to come—more than 70 percent, according to surveys conducted by Consumer Intelligence Research Partners. "For singles, it's more than just an aesthetic thing: It's a dating red flag (or green flag, as it were)," the *New York Post* warned of using Android phones.[7]

In 2013, Eddy Cue, who had negotiated Apple's book deals and oversaw digital services, began studying how to make iMessage available on Android phones, because he worried about increased competition from Google if it were to acquire WhatsApp. "We should go full speed and make this an official project," he proposed. "Google will instantly own messaging with this acquisition." But Phil Schiller, the so-called App Store chairman, balked, writing: "And since we make no money on iMessage what will be the point?"

Cue responded: "Do we want to lose one of the most important apps in a mobile environment to Google? They have search, mail, free video and growing quickly in browsers. We have the best messaging app and we should make it the industry standard. I don't know what ways we can monetize it, but it doesn't cost us a lot to run."

Craig Federighi, Apple's chief software executive, seemingly agreed with Schiller, saying he worried that opening it up as an option on Android would remove an obstacle for iPhone families to get cheaper Android phones for their children. "In the absence of a strategy to become the primary messaging service for [the] bulk of cell phone users, I am concerned the iMessage on Android would simply serve to remove [an] obstacle to iPhone families giving their kids Android phones," he wrote. Three years later, when the topic arose again, Schiller made a similar case to Cook in another email: "Moving iMessage to Android will hurt us more than help us."

Google executives, in fact, were often debating the best approach to steal away Apple customers. At one board meeting, a presentation laid out the problem: "Users are also buying into an ecosystem—not just a mobile device." Their research of customers found that Apple users with multiple devices were much less likely to switch to Android. In 2017, a group inside Google was studying the market dynamics and how they could persuade users to switch to a Google-made phone called the Pixel. A survey of users around the world returned results that showed that iPhone users were "generally more concerned than others about making the switch" and that they were "VERY attached" to key features in the Apple ecosystem, including iMessage and music, according to a report generated. Among their findings was the articulation that for users to switch, the experience was like learning a new language between the operating systems and that it wasn't easy—taking on average forty steps and as long as nine hours to transfer data. "Switching phones = switching ecosystems," Google's team concluded. Another internal study conducted later found that only 22 percent of iPhone users surveyed would consider leaving Apple while 17 percent of Android users would. Of those who actually intended to make the switch, the numbers were even smaller. Yet another internal study concluded smartphone users had an ownership bias. Android was "less trusted than iOS," the study said. "Android is less likely to be trusted than iOS and is less likely to be seen as secure."

Google didn't end up buying WhatsApp, but Facebook did—an acquisition in early 2014 that would pit the social media co-founder Mark Zuckerberg against Cook down the road in what was going to become a heated rivalry. With the smartphone central to so many lives, increasingly companies wanted their apps to be the gateway to the digital world, so-called super apps that combined communication, social media, marketing, shopping, payments, and more. Zuckerberg had many of those pieces coming together with his social media giants, Facebook and Instagram, that offered sophisticated advertising networks. WhatsApp brought a massive network of private text messaging. His teams were working to build out a market function on Facebook as well. In many ways, Apple's iPhone empire was the Super App these entrepreneurs dreamed of unseating. What made it hard for the Facebooks of the world to challenge that power was the Walled Garden. The App Store rules limited how its apps could work and collected a percentage of sales. It could be, frankly, hard to imagine what a Super App might look like in the United States given Apple's control. But one didn't have to look far to see what kind of threat a Super App might be to Apple's digital empire: China.

*   *   *

Apple only needed to look to China to see what could happen if its Walled Garden cracked and the danger that would create to its business in the United States and elsewhere. The iPhone was behind in China. It wasn't until late 2013 that Tim Cook was able to secure the kinds of deals with a key Chinese telecom that were needed to usher the device into the world's largest consumer market. But by then, a parallel world had grown up that illustrated what things could look like without Apple's dominating users' digital lives. In China, that world was controlled by a man named Pony Ma. He had become famous—or *infamous* among Western tech leaders—for his views on doing business. According to him, *to copy is not evil*. In the late 1990s, he copied AOL's instant chat feature for the China market, making him a wealthy man and positioning his company, Tencent, to take advantage of the explosion of cheap mobile phones in his home country, where a burgeoning middle class would skip computer ownership altogether for smartphones.

The company's early success with an instant messaging system called QQ would see it evolve into video games, becoming a big part of its revenue when it went public in 2004. By 2011, like other Chinese tech companies, Tencent had begun using its money to acquire companies outside China, such as Los Angeles–based Riot Games. Also in 2011, Tencent would launch Weixin, or WeChat as the version outside of China would later be known, a messaging service aimed at the emerging smartphone users. But it soon became much more. It quickly appealed to Chinese users, who largely owned underpowered smartphones compared with Western consumers. Those phones were also a lot more disposable, subject to higher turnover and lower brand loyalty. In 2014, the app launched a clever marketing campaign to grow its payment feature. To celebrate the Lunar New Year, it encouraged users to send digital red packets with money to friends instead of using the traditional red envelopes of cash. In a twenty-four-hour period, sixteen million red packets were sent, launching WeChat into the mobile-payment business. Within a few years, it would have 40 percent of the market in China.[8] To run quickly on Chinese phones, WeChat in 2017 developed a cluster of mini apps, or "mini programs," almost like Apple's App Store, that offered a growing list of services that grew more and more powerful through the network effect as WeChat reached more than one billion users. It was also, crucially, like a mini personal cloud, completely agnostic to what kind of phone you used. In this way, WeChat was the first Super App. It didn't care what phone or device you used it on, and you could access it through any device at any time. Soon, most parts of daily life in China were conducted through WeChat, from news to video games to shopping, even personal finance, helped by a large population of people who didn't use traditional brick-and-mortar banks. By then, in China, smartphone shoppers were motivated by a device's hardware capabilities, eschewing one brand for another, chasing whatever was the hot new thing. They could easily move, in part, because their digital lives were largely in WeChat, not stuck in the Apple digital empire. A survey of users found that only 50 percent of iPhone users in 2016 stayed with Apple—dramatically different from in the United States, where mostly people stayed loyal.

As Apple sales in China continued to struggle into 2017, it was clearly a threat. "Unlike the rest of the world, in China, the most important layer of the smartphone stack is not the phone's operating system. Rather, it is WeChat," Ben Thompson, the influential author of the *Stratechery* newsletter, wrote that spring. "For all intents and purposes WeChat is your phone, and to a far greater extent in China than anywhere else, your phone is everything," he added. "Naturally, WeChat works the same on iOS as it does on Android. That, by extension, means that for the day-to-day lives of Chinese there is no penalty to switching away from iPhones."

Cook and others at Apple were weary. Soon after mini programs were launched, Apple delayed an update to WeChat's iPhone app, creating tension similar to what happened at Spotify when its app updates were delayed. The disagreement centered around a tipping feature that allowed creators to accept "tips" from users. Tencent viewed the transaction between users as gifts—or "appreciation" fees—and didn't take a cut. The transaction occurred outside of Apple's payment system, which the U.S. company didn't allow within the mini programs, and looked like a loophole around the 30 percent revenue share. In April, Apple warned Tencent to shut that down, threatening to block future app updates and possible expulsion from the store if it didn't comply with its rules.[9]

During a trip to Cupertino, the Tencent team gave Cook a framed piece of Chinese folk art as well as assurance that they were no threat, reported *The Information*, a tech publication.[10] In their defense, the Tencent executives stressed the differences between tipping and in-app purchases. A détente was brokered. Tencent upgraded the tipping function to "like the author." And in September, Apple changed its App Store rules to allow users to send money gifts to other users without Apple taking a cut. But it had to be a gift, not payment for digital content or services. Tencent would soon bring back its tip function, tweaked to allow tips to be paid to individual content creators. "In the past, companies like Apple might have had a difficult time understanding China-specific features," WeChat's creator, Allen Zhang, said at a Tencent developer event. "We now all share a mutual understanding and we'll soon bring back the 'tip' function."[11] WeChat developers would figure out other loopholes in the system to avoid paying Apple's share, such as steering users to external payment systems beyond the iPhone's reach. Game developers realized that they could develop a customer service chat interface that would direct a payment link.[12]

Apple would say it applies the same standards uniformly while rival developers saw the U.S. tech giant playing favorites with a politically powerful Chinese player. Cook had to have realized how tough and different operating in China would be for the company. But at the same time, Apple executives also could have seen it as a dystopian hellscape: What if this ever happened in the United States? What if a Super App ever took that market by storm? They knew the answer. Their digital economy would falter, the iPhone would become a commodity, and Apple's flywheel would disintegrate.

And perhaps more troubling was that Tencent didn't look satisfied with staying in China. As Tencent reached market saturation there, Ma and his team began looking abroad for acquisitions, like WhatsApp, which Facebook acquired instead in 2014. It was going around Silicon Valley with a massive checkbook, making sizable investments in U.S. companies that, at first blush, seemed outside Tencent's typical swim lane. The Chinese tech giant took a 5 percent stake in Elon Musk's electric-car company, Tesla, and a 10 percent stake in the social media startup Snap. As Spotify prepared to go public, the streaming music service agreed to a stock swap with Tencent that gave the Chinese company up to a 10 percent stake. The deal, which included a cash infusion into Spotify, was seen as strengthening both companies' hands in future negotiations with music labels.[13] And Tencent made a $330 million investment into a North Carolina video game company called Epic Games.

*   *   *

Tim Sweeney's career had been about remaking his beloved company to keep up with changes in technology. He began by sharing video games he made at his parents' home, which evolved into a major game maker for consoles like Microsoft Xbox. It seemed as if Epic were on the precipice of yet another makeover. He spent $12 million to develop the hit video game *Gears of War*, which generated about $100 million in revenue for Epic—making him and the company a lot of money. The sequel was more expensive, and profits shrank. The *Gears of War 3* was even more costly to make—as much as five times the original. To do a fourth one seemed like a good way to lose money. Sweeney expected it would cost more than $100 million, and, he figured, he would be lucky to break even.[14] As Sweeney looked at the competitive landscape, the video game world was changing. Free-to-play games were becoming the rage, led by the video game maker Riot Games' *League of Legends*, that let multiple players battle each other in real time and made money by selling in-game items, mostly cosmetic changes that appeal to avid gamers. It's an economic model known in the industry as games as a service. *League of Legends* became one of the highest-grossing online games in the industry, and the Los Angeles–based company soon attracted the attention of Tencent, which acquired a 93 percent stake in 2011. The deal was struck to allow Riot to operate like an independent company. A point of tension arose when Tencent wanted Riot to release a version of the game for smartphones, where Chinese users were increasingly spending more time. When Riot didn't agree, Tencent crafted a mobile version of the game called *Honor of Kings*, which quickly became a massive hit in China with 200 million monthly users and generated $1.9 billion in 2017.[15] Eventually, Tencent bought out the remaining shares of Riot.

As Sweeney looked at the play-for-free market, he liked that Epic could control its own destiny. It wasn't designing a game for Xbox to sell; rather, it could create something for anyone with a computer—maybe even smartphones. Plus, Epic could go direct to the customer.

To transition Epic toward a free-to-play model, he turned to Tencent for help in 2012, agreeing to essentially sell 40 percent of the gaming company to the Chinese giant for $330 million. The deal valued Epic at roughly $825 million. Tencent was given two board seats. Sweeney remained the controlling shareholder. His longtime deputy and co-founder, Mark Rein, kept a small share. The success of Epic had made Sweeney and Rein even richer.

Rein used his riches in 2013 to become part owner of the National Hockey League's Carolina Hurricanes.[16] Whereas Sweeney had spent his money in the early days on flashy sports cars, his outlook on life was changing as North Carolina's richest man. He shed the cars. Instead, he was spending his free time taking long hikes in the North Carolina mountains, eventually spending hundreds of millions of dollars as Epic's value grew to preserve more than forty-five thousand acres of forest throughout his adopted home state. By 2014, he would become one of the largest private landowners in North Carolina, having jumped at opportunities to buy up land from troubled developers after the Great Recession. He looked for deals, such as foreclosed land, and then put conservation practices in place after acquiring it. "Land conservation is the one unquestionably practical and cost-effective thing we can do to protect ecology and the future habitability of the planet. You can spend $100 on land conservation and less than $5 of it is lost to overhead," he told the University of Maryland.[17] The woods reminded him of his childhood, visiting his grandmother's small farm in the mountains of Virginia. The family ended up selling the property's timber rights and eventually the land itself to pay for her healthcare bills when she grew sick.[18] It was one of those things in life that stuck with Sweeney, sparking a goal of one day being able to buy up land to preserve nature. Sweeney was even willing to fight—successfully—an electric company in North Carolina that wanted to cut across one of his properties with power lines. "My fear is that the entire forest would eventually be choked by invasive species," he told a North Carolina magazine. His wardrobe befit his outlook on life: hiking shoes and cargo pants. He still enjoyed the fast food of his youth, talking about his love of Bojangles's fried chicken. "It strikes the perfect balance between tasty food and time commitment," he told the video game reporter Sarah Needleman for a *Wall Street Journal* profile.

His business ambitions were on building, again. Tencent's investment could help fund a pivot. The Epic team had identified a small, indie game dubbed *Fortnite* that was in the early stages of development within Epic that might work as a free-to-play offering. In picking *Fortnite*, Sweeney was turning his back on the kinds of big-budget, story-driven games that had made Epic such a huge success. It wouldn't be easy. Bringing *Fortnite* out would take a circuitous route, six years in the making that at one point saw the lead designer leave as Epic wrestled with what exactly it was trying to offer with *Fortnite*.[19]

The genesis of the *Fortnite* idea had come out of internal "game jam" in 2011—a sort of brainstorming contest—where various ideas got aired, then wound its way through the creative process, settling on a cartoonish style from an original darker theme. At one point, it was seen as a game for Xbox Arcade, the online gaming service offered by Microsoft. That's the platform where Chair Entertainment—Donald and Geremy Mustard's company—had shown early prowess before turning to the iPhone for Epic's first mobile game. Then *Fortnite* was going to be for computers because Sweeney wanted to pivot the company to selling games directly to users through its website. Inspired by world-building games like *Minecraft*, the idea ballooned into a game concept known as player versus the environment where users would work in teams to harvest materials and build fort structures to fend off computer-controlled monsters that attacked at night—hence the title *Fortnite*. In July 2017, *Fortnite: Save the World* was officially released, after being out in beta, as a paid early access game with plans to make it free to play in 2019. Those plans would soon be scuttled.

After all of those years of development, Epic had come to market with the wrong kind of game. The hot thing was no longer player versus environment—players building forts to defend against game-controlled monsters—but so-called battle royal–style games made popular by *PlayerUnknown's Battlegrounds*. These were games played live online.

Inspired by the Japanese movie *Battle Royale*, the game pitted users against up to a hundred players on an island, picking up weapons and fighting to remain standing. To Sweeney's team, it was a once-a-decade moment where the industry might completely change. Instead of starting a new game from scratch, which could take months, Donald and Geremy Mustard worked to remake *Fortnite* into a battle royal game to be released in roughly three months. Donald Mustard recalled the genesis for re-tasking Fortnite came in a long car ride with Sweeney and other executives. He began writing the design document when a school bus passed their vehicle. "I'm like: 'Players are going to be on a bus in the sky and we're going to jump out of it,'" he later told journalist Stephen Totilo about the game's signature opening. They had the bones of a digital world in *Fortnite* to work with, keeping the cartoony world. They would pit a hundred players against each other in a third-person shooter. The digital world was one of a huge map that grew smaller as game play went on.

From the beginning, Epic knew the *Fortnite* remake was going to be a free-to-play game that would depend upon players buying items within the game. But the team wanted to eschew selling items within the game that gave players a better chance of winning, as other games offered. Geremy Mustard recalled: "We decided to sell only cosmetic items, things that have no value other than changing your appearance."[20] They also didn't copy a common feature in multiplayer games that allowed users to taunt rivals. Instead, they replaced taunts with funny dances.

Once the game went out in late September, the response was overwhelming—even if it was obvious to some that the game was rushed out. Internally, the team raced to build out features with a bigger vision for the digital world they had just unleashed. Or as Donald Mustard would remember it, they were building "a place where you could have all sorts of different agency driven game and entertainment experiences with your friends—all connected by your 'hub' (your locker, cosmetics, progression, etc). It was such a HUGE, audacious, vision."[21] Sweeney, too, would soon come to see *Fortnite* as something more than a game. It was a Super App of its own.

FRENEMIES

2018

The late Steve Jobs's relationship with Google's chief executive Eric Schmidt soured famously after the search giant launched Android in 2008 to compete against the iPhone. A new era of Apple and Google's relationship, however, was finding ways to work together—even as their digital empires seemingly competed for smartphone users. In 2018, Tim Cook came together with Google's CEO, Sundar Pichai, for a meeting to discuss how the companies could boost sales of Google's search business. Afterward, their deputies shared follow-up notes, including one from a Google executive to an Apple counterpart: "Our vision is that we work as if we are one company." They were already working pretty closely, which, in turn, strengthened each other's business and their own digital worlds.

Before the iPhone, Apple had a deal with Google to make it the default search engine on its computers. Since 2002, the search box in Apple's web browser was powered by Google. Around 2005, Google became worried that Yahoo might replace it in Apple's ecosystem. The search giant came to the computer company with a proposal: Google would pay Apple a onetime fee of $10 million plus 50 percent of its annual advertising revenue generated from Apple computer users' searching and clicking on paid links. In exchange, Google's search engine would be the preinstalled default search engine behind Apple's Safari web browser. The user could change the setting, but otherwise the search box was powered by Google. From the deal, Apple's revenue share totaled a little more than $100 million in 2005 and 2006. When the iPhone was launched in 2007, Phil Schiller, known as the App Store chairman, negotiated to amend the deal. At the time, Sundar Pichai, a product manager overseeing Chrome, raised concerns about the appearance of users not having a choice for search engines. "I know we are insisting on default but at the same time I think we should encourage them to have Yahoo as a choice in the pull down or some other easy option," he told the team negotiating the deal. "I don't think it is a good user experience nor the optics is great for us to be the only provider in the browser." Still, a few days later, when he saw a demo of the iPhone with Yahoo as the home page displaying its own search box, he wasn't happy. He complained to Schiller in a call, who assured him that the home page simply launched to the last page visited. Google was the default in the search box.

At first blush, setting Google as the default search function in a web browser seems like a small thing. The user can change it. But the tech world would see in coming years the power of preset defaults, especially on smartphones where the screens were smaller, and some users might struggle to find ways to make those changes. Google's own behavioral economics team would conclude, "Inertia is the path of the least resistance. People tend to stick with the status quo, as it takes more effort to make changes." Another internal study found that more than half of iPhone users in the United States were "unsure" if Google powered Safari search. The deal would be a boon for Google. In years to come, Apple-originated search traffic would add up to half of Google's total search volume. And the mere prospect of losing its special place within the Apple empire was called a "Code Red" scenario—a dynamic Apple would time and time again use to squeeze more and more money from its rival. But in 2007, before Google entered the smartphone market with its own operating system, it wasn't clear what was going to happen.

*   *   *

The first Android devices entered the market about a year after the iPhone. And quickly caught up. The South Korean phone maker Samsung would spend two years developing a wide range of phones and spending heavily on marketing to surpass the iPhone with its Android-powered devices. Suddenly Google's operating system in 2012 was being sold on more phones than Apple was selling by four to one.[1] Together, Apple and Google controlled the smartphone world, with their operating systems holding 93 percent of the global market share in 2013.[2] While Samsung might be the device, Android was the brains behind it. Apple and Android were the new Coke and Pepsi for the digital age. And Apple's Tim Cook made it clear how he saw the world when he made a rare TV appearance on the *Charlie Rose* television program in 2014 ahead of the company releasing its latest software update, iOS 8. Rose asked Cook to name his competition. Without missing a beat, Cook responded with one word: "Google."

"People would say . . ." Rose replied.

"Google," Cook interrupted.

Rose continued, ". . . Samsung instantly, because of the products. They make smartphones like this. Not like this, but they make smartphones. They have the Android operating system, which is the largest operating system in the world."

Cook disagreed: "But Google supplies that to them. And so, I think I would say . . ."

Rose continued: "Google is your competition?"

"Google is the top," Cook replied. "And then they enable many people in the hardware business—like Samsung. And Samsung is the best of the hardware companies in the Android sphere."

Rose wasn't done: What about Facebook?

"I don't consider Facebook a competitor," Cook said. "I consider Facebook a partner. We're not in the social networking business."

Rose: "And will not be?"

"We have no plans to be in the social networking area," Cook said. "We partner with both Facebook and Twitter. And we had integrated both of them into the operating system. And so we worked closely with both of them so that our customers can get access in a different and unique way to their services. And we like both companies."

They would return to the topic of Google in the context of users' personal data—a hot topic after a National Security Agency contractor named Edward Snowden in 2013 revealed government programs collecting vast amounts of data on U.S. citizens, igniting broad debate around privacy. Even before the public attention, Cook had been thinking about how Apple could compete with Google on privacy. An internal presentation circulated within Apple showed it was an issue that was top of mind. "Android is a massive tracking device," the report stated. Whereas Apple held a 2010 quotation from Steve Jobs as something of a North Star: "We take privacy extremely seriously. . . . A lot of people in the Valley think we're really old-fashioned about this."

Cook embodied that sentiment that day with Rose. "A lot of people say, you know, have said to me, there's a whole ton of information already out there, that are in the possession of companies like Google, like so many other companies, that that information is there and they worry about that," Rose said. "Too much personal information is out there and who has access to it."

"We take a very different view of this than a lot of other companies," Cook said. "Our view is, when we design a new service, we try not to collect data. So, we're not reading your email. We're not reading your iMessage. If the government laid a subpoena to get iMessage, we can't provide it. It's encrypted and we don't have a key. And so, it's sort of, the door is closed.

"But our business, Charlie, is based on selling these," Cook said, holding up an iPhone. "Our business is not based on having information about you. You're not our product."

Cook continued. "I think everyone has to ask, how do companies make their money? Follow the money. And if they're making money mainly by collecting gobs of personal data, I think you have a right to be worried. And you should really understand what's happening to that data. . . . I'm offended by lots of it. And so, I think people have a right to privacy."

*   *   *

Follow the money.

What Tim Cook didn't say on *Charlie Rose* was that in 2014, Apple was sort of getting into the web search business. The release of iOS 8 would roil Google. The software update created a new feature called Spotlight, something of a universal search function for the phone, that a user accessed by swiping down from the middle of the home screen to bring up a search field. There, the user could type what they were looking for. In turn, Apple would give suggestions, including web pages.

To build out such functions in-house, Apple that year began crawling the web, collecting data used for several things, such as improving keyboard suggestions on the iPhone. It also developed a suggestion function for Safari that allowed a user to begin typing something into a search queue and suggestions would pop—similar to how things work in a web browser. Essentially, Apple was trying to guess what the user was looking for and provide answers. Web crawling also allowed Apple to index websites so it could figure out the number of queries on a certain subject. The more web pages in an index, the better. If something is not indexed, then it is not going to be surfaced in a web search. It's used, in part, to build a knowledge graph, allowing users to ask a question and for the answer to be pulled from the knowledge graph.

These efforts sounded alarms in Mountain View, California, where Google was headquartered. An analysis prepared and shared with a senior Google executive didn't sugarcoat things. "Bottom Line: It's bad," the report began. The team expected the suggestions to siphon queries away from Google. Apple had figured out a way to both honor its contract with Google, allowing it to collect its revenue share for advertising generated when users clicked on links, and provide its own search answers in a way, Apple thought, was better for users. On the iPhone, they fretted the suggestions would "almost completely crowd out Google search suggestions" that were also presented beneath Apple's own offerings. The saving grace for Google was that Apple's suggestions weren't very sophisticated because it seemed it was often triggering irrelevant results from Wikipedia.

Apple wasn't just building search muscles of its own. It was also

entertaining new partners ahead of negotiations with Google to re-
new its multiyear contract. In 2015, Microsoft wanted to take Google's
place on the iPhone, making its Bing search engine the default. That
summer, Microsoft's chief executive, Satya Nadella, met with Cook. "I
am very optimistic that we can get a product and deal structure that
makes our two companies more competitive/benefiting users and is
long term stable," Nadella told Cook in a follow-up note. Part of the
pitch was that having an alternative to Google would help Apple in
the long run. And Nadella was so eager to make a deal that Microsoft
was willing to provide Apple with the majority of the profits generated
from the partnership. On top of that, Microsoft seemed to be responding to Apple's public posturing that it held user privacy as paramount,
offering to tighten private search for users. The offer meant just a little
under $20 billion over five years to Apple, or 90 percent of the revenue
generated. It would even raise the offer to 100 percent.

But Eddy Cue, who was brought in to cut a new search deal for Apple, was skeptical of Microsoft as a viable alternative. He knew a thing
or two about using Apple's power to his advantage in negotiations.
He'd done the book deals years earlier that got Apple in hot water
with the Justice Department. Now, as he looked things over, he held
the opinion that Microsoft wasn't as good at monetizing advertising as
Google. "If you have an inferior search engine, customers wouldn't use
it, and so, therefore, I don't know how you could monetize it well," he
would later testify. Meanwhile, if Apple simply extended its deal with
Google, it would earn $40 billion—essentially pure profit—over the
next five years and, if they reupped the deal, presumably $70 billion in
the following five years. In the previous year alone, Apple's operating
profit totaled $71 billion. The relationship with Google, which in the
beginning meant just tens of millions of dollars a year, had quietly
ballooned into a major profit center for Cook's Apple.

Microsoft's interest opened the opportunity to possibly get even
more. As Cue considered a counteroffer, he latched onto an idea that
was equivalent to asking for the moon. Apple would make the switch
if Microsoft guaranteed minimum annual revenues of $4 billion in the
first year, increasing it by $1 billion each year for the next four years
for a total of $30 billion—guaranteed. It was still $10 billion short

of what Apple thought it could easily get from Google in that time, not including down the road. It was a nice backup, though, if Google didn't want to do business anymore. In the end, though, Microsoft balked at Apple's ask.

Meanwhile, Sundar Pichai, now Google's CEO, was preparing his own pitch. Worried about the threat of Bing, Google was also putting pencil to paper to estimate what Microsoft would need to offer to win over Apple. The study, dubbed internally as Alice in Wonderland, suggested Bing would need to pay Apple 122 percent of its revenue just to equate to the 33.75 percent cut Google had been giving. The study concluded that Microsoft, "Alice," couldn't profitably match Google's offer and Apple's supposed leverage in negotiation wasn't so great after all. Plus, both sides seemingly wanted a deal. They'd push back and forth a little. Apple wanted more money. Google, worried about Apple bleeding off search traffic, pushed for a provision aimed at making sure the iPhone's suggestions didn't go any further. Both got to a yes with Cue signing the latest version of the agreement in September 2016. The terms of the deal included a provision to have Cook and Pichai meet once a year—dubbed the annual CEO check-in—to review and discuss their partnership. Increasingly wary of antitrust regulators in Europe, the contract also included a provision that the companies agreed to work together to defend the agreement if it was ever challenged legally.

*   *   *

Google was Apple's golden goose. But Apple, it seemed, always thought about how it could get more. What it needed was more leverage. In 2018, it looked as if Tim Cook found that in a man named John Giannandrea, a fifty-three-year-old Scottish native who had for eight years been a key part of integrating artificial intelligence throughout Google products, including search. That year, he became one of sixteen executives reporting directly to Cook. His hiring was called "a major coup" by *The New York Times*.[3]

He joined an organization that had been quietly putting some of the building blocks in place for its own search engine. Giannandrea's hiring would coincide with an escalation of Apple investing to develop its own search capabilities, including all of the technology needed to build a Google-like search engine. In many ways, it was like nuclear deterrence. If Apple decided to go nuclear, it would likely mean mutual assured destruction. Yet building out a full-fledged search engine would take time and money. Essentially, Apple was giving Google the opportunity to pay it not to go nuclear, to avoid unnecessary carnage. At one point, Google estimated internally that it would cost Apple something like $20 billion to catch up. Done on the cheap, maybe it could cost $10 billion to get such an effort off the ground plus $4 billion annually to pay for the technical infrastructure. Then another $7 billion annually to sustain the business with engineering and product management costs—an estimation that was conservative at best based on only a third of Google's costs. All of that would take a big chunk of Apple's R&D budget. On top of all that, Apple would be losing the cash that had been pouring in from Google.

Still, Apple wasn't short on funds. Sitting on more than $200 billion in cash, Apple had other options than just building a search engine itself. To catch up with Spotify, Apple had spent billions of dollars to jump-start its streaming music offering. Maybe it could just do that again. A brief encounter between Cook and Satya Nadella, Microsoft's CEO, sparked an idea of Apple buying Microsoft's Bing—two years after losing the bid to be the default search engine on the iPhone.

As Cook's team in 2018 tried to digest the idea, Adrian Perica, Apple's top deals guy and one of the senior executives seen as a possible successor to Cook, pressed Giannandrea on Bing's ability to help Apple compete against Google. Bing held just a sliver of market share—less than 6 percent in the United States at certain points that year—while Google dominated, generating more than $85 billion from its search business that year.[4] "What's your gut on this situation?" Perica asked. "Is the gap much larger in the user space than appears on paper?"

Giannandrea had worries. "I have been living on bing for the last few days. Mostly it works fine. Then in the odd long tail query it just doesn't," he replied. "A recent example is 'annie lennox first band.' Google gets 'the tourists' as a web answer. Bing highlights the same answer but shows a box highlighting the Eurythmics. This worries me a lot."

As they continued to mull it over, Perica wondered if Bing could help improve Siri, Apple's struggling voice assistant. "I don't think bing can do better than google search for the search use case unless it spends more on it or has a better mousetrap," Giannandrea told Perica in an email that December. Part of the problem for Microsoft was that since it didn't have a big position in the mobile space, Bing was missing out on important user data that made Google so powerful. "Not having mobile queries at scale is a huge liability for them since the most important search signal is engagement," he continued. "But it is not impossible." In theory, Apple's involvement with Bing would infuse it with the kinds of data that could jump-start it. "As we noted yesterday," Giannandrea said, "the reason a better search engine has not appeared is that it's not a VC fundable proposition even though it's a lucrative business." Though taking on Google was probably only something a major player could do, he wasn't sure it was the right path for Apple. "Can I imagine that Apple can build a search engine to compete," he asked. "Yes but it's probably not the best way to differentiate our products."

As Perica looked at the resources required, he said he liked the idea of collaborating on some projects with Microsoft. He could see a world where Siri searches, not covered by the Google contract, would be handled by Microsoft while searches with commercial intent could be conducted by Google. In doing so, it would help Microsoft get better. "We build them up, create incremental negotiating leverage to keep the take rate high from Google, and further our optionality to replace Google down the line," he told Giannandrea.

After that, Perica made a presentation for Cook that spelled out options that included growing Siri organically, collaborating with Microsoft on Siri, fully buying Bing, or forming a joint venture with Microsoft.

Neither seemed to win the day. Soon afterward, Cook and Cue met with their counterparts at Google for almost two hours. On the agenda was the political environment, including Facebook, which was increasingly at odds with Apple. It was a meeting that left the Google side optimistic. "Tim's overall message to Google was 'I imagine us as being able to be deep deep partners; deeply connected where our services end and yours begin and sees no natural impediment to us doing more together," Don Harrison, Google's president of global partnerships, reported to colleagues afterward. Harrison continued that Cook acknowledged the two companies' shared history, saying Cook "wants to figure out how we work more deeply together (and share information better—he stressed this a few times)." The relationship would continue, eventually growing so that Apple collected about $20 billion in a single year from Google, or almost 20 percent of the iPhone empire's total operating profit.

*   *   *

Mark Zuckerberg had had enough with Tim Cook. If their respective companies had been partners, as Cook called them in 2014, by 2018 they were rivals, locked in a cold war that was beginning to come into public view. After years of dancing around each other, the co-founder of Facebook was done with niceties. It felt as if Apple were picking on Facebook when it was down. Following the 2016 U.S. presidential election, Facebook had become the face of a new backlash against Big Tech. Zuckerberg was engulfed in a corporate crisis after it was revealed that a firm tied to Donald Trump's campaign had improperly accessed the data of tens of millions of Facebook users.[5] Facebook users hand over lots of personal information—not just name and hometown, but also location and likes of content. Information that's used to create a profile of the user to be targeted by advertisers. Though Facebook likes to stress it doesn't sell data directly to advertisers, it does use the user information to create tools to target ads at certain kinds of users. The revelation that an enormous amount of data collected by Facebook wasn't handled properly touched on fears that tech companies had grown too powerful. (The sort of warning Cook had been making for years about Google.) Asked during a television appearance on MSNBC before an audience in Chicago about how he would handle the situation if he were Facebook CEO, Cook curtly replied, "I wouldn't be in this situation." He continued attacking the underpinnings of Facebook's technology. "We've never believed that these detailed profiles of people—that has incredibly deep personal information that is patched together from several sources—should exist," Cook said. He advocated for government regulation. "I do think that it's time for a set of people to think deeply about what can be done here," Cook continued.

The co-host Chris Hayes pushed back. "Now, the cynic in me says, you've got other tech companies that are much more dependent on that kind of thing than Apple is. And so, yes, you want regulation here because that would essentially be a comparative advantage, that if regulation were to come in on this privacy question, the people it's going to hit harder aren't Apple," he said. "It's places like Facebook and Google."

Cook countered. "Well, the skeptic in you would be wrong," he said to laughter. "The truth is we could make a ton of money if we monetized our customer. If our customer was our product, we could make a ton of money. We've elected not to do that."

The audience cheered.

"Because we don't," Cook continued, "our products are iPhones and iPads and Macs and HomePods and the Watch, etc., and if we can convince you to buy one, we'll make a little bit of money, right? But you are not our product."

Hayes replied, "Right."

"You are our customer," Cook told him. "You are a jewel."[6]

Zuckerberg watched and just seethed. Cook sounded so glib, so self-righteous, so hypocritical. Facebook executives felt as if Cook were piling on, and as they reviewed the situation, it was clear their boss wanted blood. "We need to inflict pain," Zuckerberg told his team privately.[7] In public, he was only a little more diplomatic. "I think it's important that we don't all get Stockholm syndrome and let the companies that work hard to charge you more convince you that they actually care more about you," Zuckerberg soon told *Vox*'s Ezra Klein about Cook's comments. "Because that sounds ridiculous to me."[8]

Zuckerberg had reason to feel as if Cook wasn't being straight. On and off since 2016, the two companies had been secretly talking about ways to find common ground. Apple was pushing Facebook to change its fundamental business model, including one idea to create a subscription-based version of Facebook that would be ad-free.[9] It would be a move that would benefit Apple, of course, allowing it to collect as much as 30 percent of the fees generated. Otherwise, the ads Facebook was selling were beyond Apple's reach. Unlike Google, which had agreed to give it a cut of its ad business generated from Apple users, Facebook wasn't sharing. It was clear from their talks that Apple had its eyes on Facebook's sales, arguing that the iPhone empire was entitled to a piece of the action. Executives had zeroed in on a Facebook feature that allowed a user to pay to increase the number of people who could see a post. While Facebook considered this an ad used by small businesses to reach a wider audience, Apple argued it was an in-app purchase. In the private talks, at one point, Apple pitched working together to "build a business together."

Facebook would grow unhappy with the pace at which Apple would approve updates to its app—a complaint echoed by Spotify and Tencent—amid its tussles with the giant. In 2017, Zuckerberg tried to clear the air with Cook, meeting with him at the Allen & Co. Sun Valley Conference, a who's who of tech and media held each summer in Idaho. As they sat together, Zuckerberg raised issues about the app review delays, among other problems. In response, he felt Cook was abrasive and unwilling to give ground.[10]

Then the data controversy erupted in early 2018, and Cook's public comments became more pointed. Years earlier, Zuckerberg had pushed Facebook to develop its own smartphone, a hedge as its customers flocked to mobile from desktop. But Apple and Google had already won that battle. To do business in the digital world, Zuckerberg had to kowtow to Cook, a lesson that was going to be made painfully clear.

That summer, Apple rolled out changes to its Safari web browser under the banner of improving privacy that would hinder Facebook's ability to track users without permission as they surfed the web—valuable information that could later be used to target ads. The next step would be even more extreme. In years to come, Apple would require that all apps receive permission from users before tracking their internet usage. The policy change for the App Store was like dropping a bomb on Facebook's business, projecting to cost it more than $10 billion in sales.

Add Facebook to the growing list of those who felt as if they were living under the thumb of Apple and its Walled Garden. Something was going to have to give.

*   *   *

After spending 2018 running experiments to show how harmful Apple's actions were, Spotify was ready to do something to push back. The company's list of grievances against Apple had only grown, counting additional app update rejections and other efforts to block it from communicating with users about alternative payment methods. Chief Executive Daniel Ek's top lawyer, Horacio Gutierrez, had come up with a dual path of attack: They would seek relief from European antitrust regulators and push the European Union to pass new laws targeted at reining in Apple's power over the App Economy. The second part of the plan would take years to see through. The first part went into effect late on Wednesday, March 13, 2019, when Spotify filed its complaint with the European Commission, alleging that Apple had abused its control over which apps appear in the App Store to limit competition against its own streaming music service. It took issue with Apple blocking efforts to inform customers of ways to upgrade its service outside Apple's reach.

The following day, dressed in a black suit and white shirt, Ek stood onstage to make his case against Apple before the gathering of the world's leading antitrust regulators in Berlin for the International Conference on Competition. These were the regulators—a relatively small circle of lawyers who often work together across borders—that he was going to need to convince as Spotify took its fight to country after country. A giant screen behind him read, "Apple isn't playing fair."

"As you are aware, Apple is both the owner of the iOS platform and its App Store and a competitor to services like Spotify. In theory, this is fine," Ek told the room. "But in Apple's case, they continue to give themselves an unfair advantage at every turn—setting themselves up to be both referee and player in the world of audio streaming. This deliberately hurts Apple's competitors, like Spotify, but even more importantly, it harms consumers. I believe we are approaching an important time in history where we have to make a choice: Do we want a few, select dominant platforms to have the power to strong arm others and tax the rest of the ecosystem, taking away the ability for smaller companies to effectively compete? Or . . . do we want a healthy ecosystem where real competition flourishes and where consumer choice wins?"[11]

As if he could already sense Apple's attack, Ek attacked his rival's suggestion that if he didn't like the App Store rules, go someplace else. "As we all know, iOS and the App Store is the only way to offer our service to anyone with an iPhone or iPad," he continued. "That's over a billion people around the world. So not being on their platform is just not an option for us—or really for any competing internet service in this day and age. Apple knows this. If we wish to use Apple's payment system to allow our customers to upgrade to our Premium service, we must pay that 30% tax. This means we cannot be price competitive because we are forced to increase our cost to consumers. While Apple avoids the tax all together and can offer Apple Music at a much lower, more attractive rate. This is especially damaging to a company like ours who already pays out a significant portion of our revenues to record labels and music publishers."

Apple would strongly deny wrongdoing in a lengthy statement, essentially accusing Spotify of being a freeloader. It took issue with Spotify claiming it had blocked access to the App Store, saying it had distributed almost two hundred app updates on Spotify's behalf, allowing for more than 300 million downloads. "The only time we have requested adjustments is when Spotify has tried to sidestep the same rules that every other app follows," the company said. Apple also noted that 84 percent of the apps in the App Store were free. It argued the revenue share and in-app payments system was paying for the ecosystem that benefited everyone. "Let's be clear about what that means. Apple connects Spotify to our users. We provide the platform by which users download and update their app. We share critical software development tools to support Spotify's app building. And we built a secure payment system—no small undertaking—which allows users to have faith in in-app transactions. Spotify is asking to keep all those benefits while also retaining 100 percent of the revenue," Apple blasted. "Spotify wouldn't be the business they are today without the App Store ecosystem, but now they're leveraging their scale to avoid contributing to maintaining that ecosystem for the next generation of app entrepreneurs. We think that's wrong."

PART III

BATTLE ROYAL

10

A RISING KINGDOM

2018

With *Fortnite*, Epic Games had a huge hit. Three days into 2018, CEO Tim Sweeney was already thinking about ways to take advantage of the momentum. He asked his longtime confidant Mark Rein if he could get him a meeting with Greg Joswiak, a key deputy of Phil Schiller's, who had increasingly become a new face of Apple as vice president of marketing. Seemingly affable for an Apple executive, Joswiak might be willing to entertain what Sweeney wanted to discuss: brainstorming ways for Apple to open its Walled Garden. Given how rigorous Apple's iPhone software security was, Sweeney felt as though the tech giant could finally decouple its quality, content, and editorial decisions on the App Store from whether a user had the right to install a given program on their devices. "If the App Store were merely the premier way for consumers to install software, and not the sole way, then Apple could curate higher quality software overall, without acting as a censor on free expression and commerce on the platform that takes the form of software," Sweeney told Rein.

There was some precedent in Sweeney's mind. While Apple didn't allow typical users to download software onto the iPhone outside its App Store, for certain ones, such as major companies, it had created a program that allowed corporations to develop software to distribute directly to iPhones without being subjected to Apple's guidelines. Apple had something called an enterprise certificate that those corporations used to place their own stamp of approval onto the software for their distribution. Sweeney thought this could be a way for him to directly distribute to his users around the App Store. It might have been a clever work-around to Apple's rules. But, in basic terms, what Sweeney wanted was Apple to open up the iPhone to be just like the Mac—an idea at odds with what the late Steve Jobs envisioned and with what his deputies, including Schiller and now-CEO Tim Cook, had spent the past decade defending.

Still, Rein reached out to a contact at Apple, Tim Kirby, a strategic deals director, who seemed open to a meeting. Epic had already been working on software to bring *Fortnite* to the iPhone, which Rein dropped into the conversation, and Apple was eager to talk about it. The game was the perfect door opener for Epic. It was blowing up in the console world, and Apple would want in on the action—so much so that it was clearly willing to have a conversation face-to-face. "He was definitely receptive to the idea which doesn't mean it will go anywhere but it means he'll line up people who will listen and not shoot it down like Phil Schiller would," Rein told his friend. "He says they have had internal discussions about this sort of thing so having you in a conversation about it (even if it's a one-way conversation to start) would be useful." He added, "Maybe there's a smallest little crack on the very outer surface of the many feet of ice covering the frozen lake that is the Apple store."

Sweeney was right to think he had a unique opportunity with *Fortnite*. And he wasn't alone in seeing that Epic might have the upper hand to remake the industry.

*   *   *

By 2018, *Fortnite* was racing toward 125 million registered users that spring.[1] More than just attracting the typical gamer boys and young adults, the cartoonish battle game seemed to reach beyond the normal demographics. So much so that Steve Allison, who began his career at Atari after graduating from San Francisco State University, marveled when he overheard moms and their daughters talking about the game in public. In his more than two decades in gaming, Allison had never seen a title take over the zeitgeist like *Fortnite*. It felt bigger than a movie or TV show and definitely any other game. Little did he know, the wave of excitement was just beginning. With what he did know, Allison thought the game's success created an unusual opportunity to change the very business model of gaming. It was an idea that would two years later put Epic on a path to a very public war with Apple and Google, inserting the game maker into a geopolitical game of intrigue. Yet, at the beginning, it wasn't about opening up the App Store. It was about dethroning a different power broker.

A couple of years earlier, Allison held private conversations with Epic about the idea of creating an online store to sell video games. Though Sweeney had been experimenting with mobile games with *Infinity Blade*, the bulk of Epic's business remained on personal computers and consoles—gaming machines such as Sony's PlayStation. The company, like others, had software called a launcher that allowed it to distribute games online to computers, but it wanted more. Sweeney's team imagined a store would allow it to sell other publishers' games as well, especially those built on his Unreal Engine, the tool he designed to create video games.

The gaming industry is notorious for churn. A game is hot for a moment. Players flock to it, only to abandon it—or churn over—for something new. Games like *Fortnite* were designed to be updated regularly with new content—such as missions or maps—and features after its initial release to keep players' interest as long as possible. If players came to the game through a store, then, in theory, it might be easier to get them back if they remain in that ecosystem, keeping that churn down for all. In 2016, however, Allison didn't think Epic was ready to pull off something as audacious as opening its own digital store. Yes, it had some popular games. But they weren't *that* popular. No, Epic needed what he called an anchor, something so *hot* that it could draw people to the store, funneling gamers to other titles. That's what the video game developer Valve did when in 2003 it used the success of *Counter-Strike* to launch an online store called Steam to distribute software for computers. The quick rise of Steam to become a behemoth of online computer games coincided with brick-and-mortar retailers pulling out of the PC gaming market. Like Apple's App Store, Steam had a 70-30 revenue share with it keeping the 30 percent. At a high level, that model felt familiar to gaming companies because they typically sold their games at a wholesale price and the retailer would mark it up from there so that the publisher was essentially getting 70 percent of what the customer was ultimately paying. Except as the industry shifted to digital, there were fewer expenses involved in distribution. Just like in the book industry years earlier, there weren't all of the costs of printing the games on disks and distribution. But if a game publisher didn't like the math, then tough luck. Only a few games with mass appeal had demonstrated an ability to bypass Steam. They were blockbusters, games such as Riot's *League of Legends* and Mojang Studios' *Minecraft*. Most publishers, including Epic, *needed* Steam. But, in Allison's opinion, Steam was vulnerable to competition, and *Fortnite* could provide them with the perfect weapon for battle. That spring, Allison dashed off an email to one of Sweeney's top deputies, Paul Meegan. "Dropping you this note out of the blue to say I'm so excited for you guys and how Fortnite is blowing up pop culture," he wrote. "I've been in places like Panera (last weekend) here in Walnut Creek [California] and overheard a table full of moms and 12–14 daughters talking about Fortnite—and it made me smile because that shows the reach."

He continued by saying that he thought Epic was on the cusp of something even bigger. "Fortnite blowing up definitely has created that potential Valve-Counterstrike moment at a scale that is much bigger than when that gave birth to what is now Steam and their 80 to 95% share of the entire PC digital market (depending on genre)," Allison wrote. By his math, Valve's Steam was taking a 30 percent cut on 85 to 95 percent of the paid PC game market, "fairly effortlessly," allowing them to bring in more than $4 billion annually. "Steam is pretty ripe for disruption—if [you] wanted to take 20 to 30% paid digital PC market share you could like nobody else can," he wrote. "It doesn't take much to get the rest of the PC development and publishing world to want to look to a new focused solution more because of the flood of titles to Steam that is creating that App Store like issue of discovery and regressing per title revenue than the 70/30 model."

He figured a revenue share of 80-20—with Epic keeping 20

percent—or maybe even 90-10 or 95-5 for games based on Epic's gaming engine would attract attention. Such an attractive revenue
share "paired with the scale you've just hit would immediately disrupt
the entire industry," he concluded.

The next day, a Saturday, Meegan forwarded Allison's email to
Sweeney, the CEO's longtime confidant Mark Rein, and some others inside Epic Games. Allison had their attention. What he had no
way of knowing was that Sweeney and his team were already thinking
about ways to expand their digital kingdom. A few weeks later, they
hired Allison to help build an online store. Soon, though, they would
learn a painful lesson: Taking on Steam might be one thing, but taking
on the real kings of video games would be another.

*   *   *

The economics of *Fortnite* were a world away from the early days of
video games when players paid a quarter to play *Pac-Man* at the arcade, or $25 for *Super Mario Bros.* to play on a Nintendo machine at
home in 1985.[2] The reimagined version of *Fortnite* that offered a mode
of play called *Battle Royale* was free to play, pitting players against each
other until only one person or team remained. Epic made money by
selling digital things within the game using a virtual currency dubbed
V-Buck, which players used real dollars to acquire. V-Bucks could then
be purchased to acquire add-ons in the game, such as outfits to make
their avatars appear like Marvel Comic superheroes. Since only a small
percentage of users actually spent, Epic's strategy was to make the pool
of players as big as possible, fueling a push to allow players to compete
beyond one single console or PC and play against players around the
world using their favorite device. A game that was platform agnostic, a digital kingdom accessible through an Xbox or an iPhone or
whatever—where everyone was an equal.

Tim Sweeney wasn't the first to think up the idea of cross-platform
play. However, the game machine makers Sony, Nintendo, and Microsoft weren't united on supporting the idea—for understandable
reasons. They had built worlds of their own. Some in the industry saw
potential, but the biggest one resisted. Sony, maker of the popular

PlayStation, had objected to efforts by Microsoft to have the popular *Minecraft* game play across Xbox, PlayStation, and Nintendo's Switch. In 2017, Xbox and Nintendo began allowing cross-platform play of *Minecraft* to the chagrin of PlayStation players. Around the same time, *Fortnite* players during a weekend in September began to notice something weird about the game. A PlayStation player noticed he was eliminated in the game by a username with letter configurations disallowed in the Sony system. He rooted around a little, only to realize it was a user playing via Xbox. The unthinkable was occurring: cross play between PlayStation and Xbox! He posted his discovery to Reddit, the social media site, and it quickly took off among gamers.[3] By Monday, Epic confirmed it had "a configuration issue and it has now been corrected." The response among gamers was a mixture of intrigue and outrage. "Since allowing cross-console play seems to be so easy that a company like Epic can do it by mistake, lots of players are now asking why the hell there are so many roadblocks to getting the feature," Cecilia D'Anastasio, a reporter for the video game website Kotaku, wrote.

Behind the scenes, Sweeney was working quickly to try to take advantage of the increasing interest in *Fortnite* and strike deals advantageous to his blockbuster. Microsoft and Nintendo seemed willing to allow cross play, and his team was working to expand on Apple and Google's Android. Sweeney wanted to expand his digital kingdom.

*   *   *

In January 2018, a team from Microsoft visited Epic's headquarters in North Carolina to discuss how to make cross-platform play work. Back in Redmond, Washington, Microsoft's Xbox boss, Phil Spencer, had some concerns. Sweeney was planning to launch *Fortnite* on Apple and Android devices in late February, supporting cross play, including allowing cross purchases and cross progression. In other words, V-bucks bought on one platform would count elsewhere, and players could stop play on an iPhone and pick up play again on another platform, like Xbox.

Sweeney tried to assure him. "It's bringing together current and potential gamers in real-world social groups: college dorms, high school classes, even kids, as only Minecraft has done so far," Swee-ney told Spencer. "We want to work with Microsoft to unblock all console-mandated interop restrictions in time for this launch." And, in Sweeney's increasingly bold style, he cautioned that he thought deals with others were possible. "Platforms that block interop will be siloed," he warned. By his calculations, it was good for Microsoft to allow cross play because *Fortnite* going on to Apple and Android devices would be a "game-changing" experience. "It will sell lots of Xboxes as players are pulled into the small-screen experience socially, and want to upgrade to the much better big-screen experience," Sweeney told him. "The good this will do for console gaming as a whole will far outweigh any zero-sum effects of cross-console purchase portability."

On the whole, Spencer responded positively, though a little irked at the rushed timeline that Sweeney was springing on him with only a few weeks' notice. He was most concerned about the economics of cross purchase. Microsoft and other game machine makers heavily incentivized the sale of their machines in a bid to ultimately make their money from software and in-game purchases over the life of that machine. He also worried about items that sold in *Fortnite* played on Xbox not being competitively priced to other platforms. "One bet is every person just 'buys' on device they primarily play on," he told Sweeney. "I think that's a reasonable bet but the downside of that not being the case is a lot larger for a console than a phone or a PC." Less elegantly put, Xbox had a helluva lot more to lose than Apple.

Sweeney tried to assure him that *Fortnite* looked best on TV, not some smartphone, which he saw as a gateway for bringing in new players to consoles. The industry's problem, in his opinion, was that while mobile gaming was growing fast in popularity—its base grew five times—it failed to help the console industry because the kinds of games that were popular were for casual play on an iPhone and were disconnected from the elaborate ones played in the PC and console world. With *Fortnite*, Sweeney was laying out a way to bridge the mo-bile crowd into the console world—just when the industry needed it most. He was promising to bring console-quality games to the smart-phone without compromising. "Console and PC do need an influx of new gamers as the older generation ages out," he wrote. "All adults have smartphones and kids have hand-me-down devices. Once they've onboarded to Fortnite there, the Xbox proposition becomes: upgrade and you can have a vastly better experience, play a game you already love, and bring your stuff with you."

All of that, of course, depended on having *Fortnite* on Apple and Android devices.

*   *   *

In early 2018, the second part of Tim Sweeney's plan to expand the Epic digital empire began to take root. In his mind, initially, he was thinking of taking *Fortnite* to Apple first, then Android one or two months later once Epic was able to handle the added scale required for so many mobile users. But then, rather than just simply distribute the game through Google's app store, Sweeney had developed another idea. "What is the feasibility of launching Fortnite on Android as a stand-alone installable program, avoiding Google Play and their 30% tax?" Sweeney asked his senior team in late February. He imagined the game's website could have a download button for Android just like on computers that download the game installer along with instructions for how to load a game outside the Google app store, a rarely used process called side loading. "This could get us to having a multiplatform (PC/Mac/Android) ecosystem much faster than other avenues," he continued.

After that, he could envision negotiating bundling deals with the major smartphone makers, all of which could lead to Epic having a foothold on Android devices outside Google's grasp, allowing him to create a multi-platform app store for third-party games. "This is exactly the process Tencent followed to bypass Google Play with WeChat, which they soon opened up as a game distribution platform," Sweeney told the group. "The sooner we can free ourselves from the App Store distribution monopolies, the better, and the Fortnite launch on Android seems to be the one moment in time when we have sufficient gamer excitement to launch successfully and build a huge gamer base."

When it came to Apple, Sweeney didn't see a way for users to download apps outside its App Store. Daniel Vogel, Epic's chief operating officer, loved the Android idea and agreed that Epic had a rare moment to strike while the iron was hot. "I don't think we are ever going to have this amount of leverage again to get players [to] jump through hoops to install our 'store'/ecosystem/launcher." Another benefit, he added, was that by launching with Apple first, Epic would be able to focus its engineering efforts there to speed up launch plans while also negotiating extra support from Apple for the exclusive mobile access. All the while, he suspected, they would be building demand among Android users, which would help them negotiate deals with Android phone makers. "At some point the DOJ is going to break Apple's stranglehold on the app store," he concluded.

As the group grew more enthused, a proof-of-concept launcher was created that allowed them to download and install *Fortnite* on Android phones. The bells and whistles weren't there yet, but it worked, providing them with something to play with internally. Sweeney grew excited, too. He liked the idea of getting *Fortnite* on phones, then their own app store later. He wanted to begin negotiating with major phone makers, such as Samsung. But he cautioned his team to stay below the radar. "Here is our official plan for communicating with Google about bypassing the Google Play Store: SAY NOTHING TILL IT SHIPS," Sweeney told them.

*   *   *

Pressure by Apple's chief executive Tim Cook to squeeze more out of its digital services business was clearly evident in the success of the App Store. But the App Store of 2018 was nothing like the early days of fart apps and other amusements; it was big business for the gaming world. Apple's fee from the App Store would generate an estimated $9.47 billion from gaming that year, or roughly 70 percent of all the money from its digital market.[4] On a revenue basis, that made Apple a smaller player in the video game world. Microsoft's bean counters privately estimated Tencent was the largest player by sales with revenue. But because Apple's App Store sales were mostly pure profit, the results to the bottom line were enormous. An internal report given to Cook calculated the store in the 2018 budget year was generating a 74.9 percent operating margin—a measure of profitability that subtracts cost from revenue. That would imply an operating profit that year of $7 billion, according to an analysis, or roughly 10 percent of its $71 billion operating income that fiscal year. Another way of looking at it: Apple in that twelve-month period profited more from video games than Xbox, Nintendo, PlayStation, and the game maker Activision Blizzard—*combined*. Simply put: Tim Sweeney needed to be on the iPhone if he was going to continue to fuel *Fortnite* growth. With Xbox, Nintendo, and PlayStation, he was trying to build a unified world where he could create the biggest pool of users and, perhaps more important, *spenders*. The obvious next step was Apple. In particular, what made an iPhone-*Fortnite* pairing appealing—even if Sweeney had objections to its App Store rules—was the kinds of sales being conducted in the Apple ecosystem. *Fortnite* depended on in-app purchases for revenue, and the iPhone empire had some real whales—users who spent big. In the previous year, for example, almost all—88 percent—of App Store game billings came from just 6 percent of its customers who, on average, spent more than $750. The biggest ones, the real whales who made up 1 percent of Apple gamers, generated 64 percent of the billings with their average spend of $2,649.[5]

As appealing as the iPhone empire was to Sweeney, Apple itself was eager to have *Fortnite*—the hottest game of the moment—available through the App Store. "Fortnite: Battle Royale is coming to mobile in mid-March and has the potential to be the biggest game on iOS," Shaan Pruden, Apple developer relations executive, told Trystan Kosmynka, senior director for the app review team, in an email. "App Store is very excited about this release and is prepared to get behind it in a big way." But Apple wanted to exclusively release it for a limited time before Android. In turn, Sweeney's team pushed for the app review process to go faster. They demanded a two-hour turnaround. His long-term ask was that Apple create some sort of "trusted developer" program to expedite updates to the app, but in the short term he was pushing for a forty-eight-hour turnaround. Even with that, Epic was concerned that Apple was going to bottleneck its entire ecosystem of updates. "They have to update every platform simultaneously or the cross-platform matchmaking breaks," Pruden told the App Store review boss. "Sony and Microsoft have guaranteed a 2–4 hour [agreement].

With a weekly content update cadence, getting stuck in App Review is their single biggest concern/risk."

Sweeney's concern about app review turnaround times was as old as the App Store. Even with all of the improvements made in recent years, it could still be a cumbersome process that left outsiders fretting. And Kosmynka was skeptical of the ask. "This won't work [the] way they seem to think it will work, regardless of how quickly a review occurs," he told Pruden. "It can take several hours for App Review to even receive an app, it's not the norm but it does happen. . . . After an approval it can take up to several hours for a new app or update to make it through the . . . pipelines to customers."

The role of helping Epic do that fell on Michael Schmid, who had joined Apple's App Store for business development a few months earlier after a career in the mobile gaming industry. He was tasked with helping them get prepared for the App Store launch, working to make sure their storefront was the best it could be, and taking care of back-end issues that arose. It was a "never-ending crescendo of support," he would later say in court records. "It was quite intense." There were phone calls and text messages at all times of the day—3:00 a.m., 5:00 a.m., Christmas. He assigned someone in Australia to provide Apple with twenty-four-hour coverage. "There was a lot of good times and a lot of really stressful times," he said. "I would . . . overall describe it as tumultuous." He saw it as his job to help Epic navigate Apple's rules to ensure the game was available for users. "I came from mobile game development. I understand and I emphasize with our developers, especially those that are kind of thriving on live ops— . . . the lifeblood of your game is bringing new content to your game on a very frequent basis," he said. "*Fortnite* did this incredibly well." But because it was so focused on bringing out new content, he said, Epic often didn't have the bandwidth to step back and fix systemic issues, such as planning around submissions to the App Store. This resulted in many instances of Schmid stepping in to help Epic expedite requests for App Store review. More than just technical help, Apple said, Epic wanted rules changed to benefit it. In the first year on the App Store, Epic asked for—and Apple agreed to—an exemption from Apple's rules against in-app gifting—a rule that had been put in place to avoid potential fraud. According to Schmid's telling, Epic seemed quick to remind Apple that with *Fortnite* being such a big hit, the developer had leverage of its own. "On a variety of occasions, Epic personnel have told me that if Apple did not comply with its demands, Epic would simply terminate its relationship with Apple and remove its games off the iOS platform," Schmid said. "Epic has repeatedly told me that it could do this because Apple is the 'smallest piece of the pie' when it came to Fortnite revenue."

*   *   *

The pieces in Tim Sweeney's complicated puzzle were coming together. Epic was going to launch on Apple first while pushing back the launch on Android. Microsoft and Nintendo were on board for allowing cross play. The last impediment to Sweeney's vision of cross-platform play sat in Japan, where Sony leaders objected to the idea. Sony executives worried allowing cross play would expose its consumer-behavior data and give rivals an unfair share of their business. Increasingly, they felt Sweeney was backing them into a corner, painting them as the bad guys among gamers, who have a reputation for being somewhat fickle. Social media hashtags #blamesony and #notfortheplayers were popping up to their dismay. Sony's reluctance roiled Sweeney, who saw it as an existential threat. "Sony's policy was unfair to gamers, it was unfair to game developers, and it was ruining gaming at the time," Sweeney would later say. "Because Fortnite was going to become this mass phenomenon, the No. 1 game in the world, the inability of players to play together was a critical limit, not only to Epic's ability to build our business and satisfy our customers, but on all game developers to do the same thing."

His team pressured Sony. Joe Kreiner, a senior business development executive at Epic, offered up several sweeteners, such as committing a game to the launch of Sony's virtual reality platform, sharing marketing data, extending certain software licenses with favorable terms. "We love working with PlayStation, and we want this to be a win/win," Kreiner told Gio Corsi, a Sony executive. "The longer this drags out, it will be less so. I can't think of a scenario where Epic doesn't get what we want—that possibility went out the door when Fortnite became the biggest game on PlayStation."

Corsi disagreed—even if *Fortnite* was a hit. "Many companies are exploring this idea and not a single one can explain how cross-console play improves the PlayStation business," he responded.

Finally, after months of discussion, in a prelude to things to come with Apple, Sweeney threw down what appeared to be a final gauntlet to Sony, threatening to use its power to update its software to unilaterally allow *Fortnite* players across platforms to compete, including Sony, with or without the Japanese company's permission. Such a move would be a major provocation. Sweeney delivered the ultimatum in an email to Phil Rosenberg, Sony's U.S. executive in charge of managing relationships with game makers. Sweeney told him to let Sony's video game chief executive, Tsuyoshi Kodera, in Japan know. "Please inform Kodera-san, and please be clear, that Epic will enable full interoperability between all platforms in Fortnite at a timely point in the future," Sweeney told Sony's gaming man in the United States. "Please understand that Epic firmly believes that this action is a matter of legal right and moral duty, and that I am conveying this to you following extensive deliberation and legal analysis across several continents with Epic's board members and major investors, and that we are prepared to pursue this course with all available resources, wherever it leads us, and for however long." Or, in other words, don't fuck with us. That message was delivered to Japan, and Sony appeared to blink. A verbal agreement was soon reached. They would agree to open up cross-platform play to not just *Fortnite* but all developers.

Despite that, Sony refused to allow cross-wallet function, meaning V-Bucks bought in the PlayStation store couldn't be used outside PlayStation, and similarly, those bought outside the platform couldn't be used inside it. And quietly, Sony and Sweeney came to an unusual agreement—one Epic had not given to any other partner: He agreed to pay Sony compensation to enable cross play. If a user was primarily playing on PlayStation but making in-app purchases using an iPhone, for example, then this could trigger Epic paying Sony a revenue share to offset any reduction in revenue the console company might see by allowing cross play. It was the kind of secret deal Sweeney would later come to object to when trying to build his business in the mobile space. It was also the kind of hard-nosed tactic that Apple wouldn't tolerate.

11

SWEENEY'S REBELLION

2018

The summer of 2018, in a somewhat lengthy email, a manifesto of sorts, Tim Sweeney dropped a bomb on Google. In short, the Epic Games chief executive had decided he wasn't going to use Google's app store, Play, to release *Fortnite*. Instead, he would make the block-buster available for Android users by download from Epic's website. From there, users would have to navigate the cumbersome process of so-called side loading the game onto their devices. The move, Sweeney wrote, was motivated by a belief in "open platforms" and showing that the dominant app stores weren't adding value commensurate with the 30 percent fees they collected. Android's origin, after all, was that of being open and free. And Sweeney referred to that legacy, calling Android a great open platform like Windows and Mac (or, at least, that's how it had marketed itself). "Being open brings robust competitive and economic benefits," he wrote. "It doesn't equate to malware vulnerability as Apple touts."

The email arrived in the inbox of Jamie Rosenberg, whose job for the past eight years was heading the Play store, Google's answer to Apple's App Store and an evolution from its Android Market. Its original app store claimed its goal was not about turning a profit.

But such ideals had long ago been abandoned. From the beginning, Rosenberg was playing catch-up trying to compete with Apple's App Store, a position that had grown more challenging in recent years as the iPhone evolved into a video game machine. Google had grown accustomed to hot apps launching first on Apple and then on Android, but the prospect of Google missing out entirely on the hottest game was bad. Worse was the precedent that Epic would be setting. That was *terrifying*. His mind raced through the domino effects it might trigger, the contagion, as other big game makers saw what was possible and tried it themselves. Sweeney might not just crack open Google's digital ecosystem; his actions threatened to tear it down entirely. Shocked, Rosenberg typed out a request for a meeting in person. "As you can imagine there are surprises for us in here, and we have a number of questions," he responded.

Sweeney gave little ground, responding that he would be available the following week for a video call. And, he added, Epic wasn't seeking a counterproposal or special terms. A line in the sand had been drawn. "We believe," Sweeney concluded, "the future of computing relies on open platforms, and without robust open platforms and real consumer choice, the tech industry will continue to be transformed into some sort of cable TV style intermediation hellscape, which isn't in Epic's interest nor perhaps in the overall interest of Google."

If Sweeney suspected Google would respond with a special deal to bring *Fortnite* to the Play store, he was correct. Rosenberg quickly began marshaling resources to come up with a plan to get *Fortnite* on Play. His team ran the numbers, calculating that the absence of the game on the Google app store could result in as much as $250 million in lost revenue. But the bigger effect could be the loss if other developers followed suit. He was looking at a potential loss of $3.6 billion. On top of that, they worried about the public relations nightmare that would be created questioning the value of Play as a platform for gaming, hurting the users' perception that its store was "the best source" for apps and games.

Google had no interest in changing the rules of its app store to appease Sweeney. Instead, Rosenberg began crafting a proposal so lucrative that he hoped it would persuade Sweeney to change his mind.

A few weeks after Sweeney's email, Rosenberg took his plan to Google's powerful Business Council, which included Chief Financial Officer Ruth Porat and was responsible for approving certain projects that required substantial spending. What Rosenberg had in mind was substantial. And complicated. Rather than just hand Sweeney a pile of money, he turned to Google's traditional playbook of pulling the company's various levers to offer up value. Deals with YouTube, cloud computing, ad partnerships. In total, his team estimated the offer was worth $208 million to Epic over the course of three years. He wanted to assign twenty-four full-time employees to work with Epic, too. The cost to Google was effectively more than it anticipated receiving from its share of the revenue generated on the game if it were on the platform. "The reason to be willing to provide value greater than expected rev share is to hedge against further downside risk of contagion to the next likely set of developers but not breach current economics thereby limiting the investment proposal," a member of the Play finance team told Google's chief business officer, Philipp Schindler.* Porat, the CFO, approved the plan.

Google was willing to throw money at the problem to fix it— a tactic that was not without risk. If word got out, they could find themselves in difficult conversations with app makers. In particular, it could undermine relationships Google had forged with other key video game makers. Some inside Google were especially worried that the terms of Epic's deal would leak to other game developers, who would want similar deals. They were especially aware of the fact that Epic shared Tencent as an investor with other game makers, such as Activision Blizzard. But as Rosenberg's team laid out in a slide deck to the Business Council, they had little choice against the possible Epic defection: "This creates significant risks for our Android ecosystem & Play business, and threatens our business in the long term." As they thought about the long-term ramifications, they also began to worry that Epic's move might legitimize the phone maker Samsung's own

---

*The team also proposed an even bigger package for Epic, one valued at almost $500 million that would have included launching *Fortnite* on a gaming console Google was secretly working on called Yeti.

nascent app store and increase a perception gap of Android devices versus iPhone abilities among smartphone buyers.

While Rosenberg persuaded the leadership group to approve the plan, he'd have less luck with Sweeney. Rosenberg was able to talk Sweeney into a face-to-face meeting in Mountain View, California, where the tech giant was headquartered. But once there, Sweeney couldn't be persuaded. "It seemed like a crooked arrangement to me because, you know, our complaint at the core was that Google charged far too much for Google Play distribution," Sweeney would later recall. "Google was proposing a series of side deals which seemed designed to convince Epic to not compete with them in stores, not compete with them in distributing our software directly, and to make it look to the outside world like we bought into their 30% terms when, in fact, we were very much in disagreement."

On the way out, Rosenberg's boss, Hiroshi Lockheimer, left Sweeney with the perception that Google wasn't done fighting back and would ultimately meddle with their plans to launch that August.

*   *   *

Even before Epic freaked Google out, the Play store boss, Jamie Rosenberg, was growing concerned about the tech giant's ability to compete with Apple in offering video games. Developers were giving the iPhone exclusive first access to hot games. Samsung, too, was worried as arguably the biggest smartphone competitor against Apple. The South Korean phone maker ran on Google's software, while the iPhone ran on Apple's software. Executives from Samsung and Google had met around the time of Silicon Valley's annual developer conference in the spring of 2017 to discuss how they could improve things. But, as 2018 unrolled, Rosenberg noticed signs that seemed to suggest Samsung had bigger ambitions for its own app store, the Galaxy Store. He began hearing that Samsung was talking to top game developers about using Samsung's app store for distributing their games. He had also noticed Samsung give its Galaxy Store a default spot on the home screen alongside the Play store. Then, in June of that year, Epic privately dropped its bomb about planning to distribute

Fortnite outside Play. Rosenberg's team warned the Google brass that such a move could help Samsung but not in the way they first thought.

When his team learned that Tim Sweeney was going to be at Samsung's annual "Unpacked" event, where the company revealed its flagship smartphone similar to how Apple juiced interest each year for its latest iPhone, there was something of a panic in Mountain View. Emails flew back and forth as everyone scrambled to figure out what was going on and how bad the damage would be. The Google side had already prepared a public relations response to the idea of *Fortnite* being launched outside Play. "We continue to work the media on background and coverage is still mixed, with journalists enjoying the idea of Fortnite on Android and Tim's crusade but noting the challenges and security risks being off-Play creates for users," Colin Smith, a longtime tech PR man, wrote to the team the day after Sweeney's Android plans were announced.

But in the overnight hours, Rosenberg warned everyone that he had heard Sweeney was going to be at the Samsung event the next day. Details were sparse. From what he was hearing, Sweeney was going to talk about Epic's collaboration with Samsung to optimize performance of *Fortnite* on the new Note 9 smartphone. "He will announce that Fortnite is available for Note 9 and the Galaxy 10 S4 tablet—*in beta*," he warned the group. "No plans to discuss the distribution model on stage (per Samsung)." Furthermore, he understood that *Fortnite* was going to be distributed through Epic's website, though he seemed assured that Samsung wouldn't have a part in launching the app. "Waiting for double-confirmation from Samsung that they will not be hosting the installer . . . in Samsung app store," Rosenberg wrote.

While they waited, Rosenberg's team mobilized to shore up support among top game developers, in particular Tencent portfolio companies, by throwing "extra love" and promotion their way. "We want these folks to have a very strong Q3 . . . in particular if and when comparisons start being drawn to off-Play performance," he told his bosses. "Longer term, the team is working on a proposal for a more formalized cross-Google package of benefits and services for game developers. The Epic/BC exercise was instructive in showing what we could potentially put together . . . if we were focused on this across the company."

*   *   *

In early August, Samsung gathered journalists and others at the Barclays Center in Brooklyn for an event to reveal its new flagship phone, the Galaxy Note 9. In previous years, the tech company had pushed the limits of features in bids to outshine Apple with the first plus-sized smartphone and iris scanners to unlock home screens. This year, however, was a ho-hum offering. What the Note 9 did offer was computing power to rival a laptop, an important feature for gamers in particular. Landing the CEO of the hottest video game as a speaker at the event was a win for Samsung. About twenty minutes into the event, Tim Sweeney was welcomed onstage to excited cheers. Standing beneath a giant video screen, he wore an ill-fitting purple T-shirt that read "Fortnite."

"Judging by the tweets," Sweeney told the crowd, "just about every gamer wants to know the same thing: When is 'Fortnite' coming to Android?"

He paused awkwardly before continuing in a monotone voice: "So, about that . . . We're going to be launching 'Fortnite' beta on Android this week, and players with Samsung Galaxy devices are the first to be invited, starting right now."

More cheers.

What wasn't discussed onstage was that Epic had worked out two ways for Samsung customers to download *Fortnite*. They could side load, or they could use Samsung's app store. While Epic was fighting Google over its 30 percent cut, the game maker quietly negotiated a better deal from Samsung, which agreed to take just 12 percent on in-app purchases of *Fortnite* downloaded through its app store. That was dramatically lower than its normal 30 percent and the 20 percent Google suspected it was offering other special partners. After the Samsung event, Sweeney followed up with the phone maker's executives, including DJ Koh, the chief executive of the company's mobile business, to thank them. The Korean company had taken a big risk in helping Epic poke Google in the eye, and Sweeney seemed to acknowledge as much. "You have my assurance that Epic will support Samsung 100% in any battle with Google," he told the men. "We will not ever give in to Google pressure to support Google Play, even if Google blocks Fortnite on Android, and even if the battle requires litigation lasting many years."

He continued that the best defense was the "love" customers had for *Fortnite* and Samsung. "Though Google has threatened us in private, they are unlikely to actually retaliate, as that would be visible to all of our customers, the industry, press, and regulators. We'll stand firm in our partnership to advance the whole industry together!"

The feelings appeared mutual. "We at Samsung will do the best we can to make this collaboration a game changing moment," Thomas Ko, a Samsung executive, replied. "Entire industry has heard and now they are watching us."

As *Fortnite* went live, the Google executives seemed glued to their phones. "Opened my note8 today, went to galaxy apps, saw fortnite banner so I gave it a try," Sameer Samat, Jamie Rosenberg's colleague, told his colleagues. "Something is wrong. The Epic installer never asked me to turn on [unknown sources]. In fact, I checked after it was done and it did not have the permission. But it was able to download and install Fortnite. I have Fortnite now on my Note8 running?!"

A colleague wrote back theorizing that Epic may be leveraging the Samsung app store's permission to install apps. "That would be a clever way for them to avoid the unknown sources friction entirely," he wrote. "Well," Samat told the group, "that seems quite bad."

Meanwhile, Rosenberg, who had been assured by Samsung that *Fortnite* would be allowed only on select devices, dashed off an angry text to Jay Kim, a senior executive at Samsung, that read as if he were stinging from betrayal: "Someone on our side was just able to fully download Fortnite on Note 8 with unknown sources. We really need to understand what's going on (and I think DJ should, too). Very concerned. Also surprised that it's on Note 8 given what you said about Note 9 and Tab S4 only."

Soon Rosenberg received a text from an underling of Kim's essentially trying to excuse away what happened: "It was done by the service team without my knowledge. I am looking into it now."

But the damage was done. If Rosenberg and his colleagues were worried that Epic's action would spark other game developers to follow suit, they were only realizing the full scale of threat—a full-blown rebellion. Samsung had just demonstrated it could—and would—launch games outside the Play store. Google had nothing short of a crisis at hand.

*   *   *

Google quickly launched what it dubbed the Fortnite Task Force. Their goal was to "quantify and protect" Android users who installed Epic's game. To build off the concerns first noted when they watched *Fortnite* go live on Samsung phones, the security team worried Android users would be vulnerable to hackers finding vulnerabilities created by the way Epic and Samsung had set up their release. They soon discovered that the file format used to distribute and install Epic's software had a flaw that hackers could exploit. "This vulnerability allows an app on the device to hijack the Fortnite Installer to instead install a fake [file] with any permissions that would normally require user disclosure," Google warned Epic in a message known as a bug report. Normally, such disclosures are made public after ninety days or shortly after a patch has been made broadly.

Privately, the task force debated its next steps, according to notes of the meetings. The group wanted to get word out of the bug quickly. Sameer Samat, the Google executive, tried to frame the issue as such: They wanted Samsung to not allow *Fortnite* to be installed outside app stores and for other developers to realize following such a path would be messy. "On Samsung," he said, according to notes, "what is the best way to make them feel a tremendous amount of heat?" He wanted the task force to figure out a way to escalate the matter to a conversation between their boss and their boss's Samsung equivalent, delivering the phone maker a message that "you're just creating vulnerabilities and we're both going to end up losing."

Once the bug was flagged, Epic jumped on the problem immediately, rushing a fix within about a day and asking for ninety days before going public with the issue so the update could be widely installed. But inside Google, executives wanted to go faster, deciding to release the info a week later while plotting how to make sure Epic's misstep got widespread media coverage. Details of the bug soon found its way into news reports. "This was exactly the kind of exploit that Android Central, and others, had feared would occur with this sort of installation system," Andrew Martonik of the enthusiast website Android Central wrote.

Sweeney responded with a statement that was critical of Google for rushing to disclose the bug. "A company as powerful as Google should practice more responsible disclosure timing than this, and not endanger users in the course of its counter-PR efforts against Epic's distribution of Fortnite outside of Google Play," he said.[1]

*   *   *

Winning over Samsung was just the first step for Tim Sweeney, who had designs on creating app stores for not just Android but also Apple. To do so, however, would require even more might than just *Fortnite* could bring.

Late that year, Sweeney approached his rivals at Valve to encourage them to abandon their 30 percent commission on their online game store and adopt a similar revenue share as Epic was rolling out for its online store. He worried that a proliferation of backroom deals, such as the kind he was offered by Google, for major game developers would torpedo any effort to shame Apple into abandoning its 30 percent commission and what that might mean for the next generation of technology—whether that be augmented or virtual reality or artificial intelligence. In a way, Valve's own 30 percent commission helped legitimize what Apple was doing. In an email, Sweeney noted that the Supreme Court was hearing a case that could allow consumers to file antitrust lawsuits against Apple for alleged damages caused by its App Store rules. It seemed like momentum was building for Apple's Walled Garden to be properly challenged in court. "However, Apple also has the resources to litigate and delay any change for years, which will cause their iOS monopoly to spill over into an AR monopoly that's emulated by Google, Microsoft, and others," Sweeney wrote. "What we need right now is enough developer, press, and platform momentum to steer Apple towards fully opening up iOS sooner rather than later. If Apple opens up iOS, then all 5 of the world's top general computing platforms will be open, and it will be very hard for any new platform to change the status quo."

Valve didn't seem interested in joining Sweeney's rebellion. "Right now," Sweeney followed up a few days later, "you assholes are telling the world that the strong and powerful get special terms, while 30% is for the little people. We're all in for a prolonged battle if Apple tries to keep their monopoly and 30% by cutting backroom deals with big publishers to keep them quiet. Why not give ALL developers a better deal? What better way is there to convince Apple quickly that their model is now totally untenable?" If anything, the Valve executives seemed bemused by Sweeney. One executive wrote the others internally to ask about Sweeney's sharp words, "You mad bro?"

But Sweeney wasn't to be deterred. By the spring of 2019, Sweeney was pitching the idea of an industry app store. He was trying to convince Ubisoft, Supercell, and Activision Blizzard. He proposed to Armin Zerza, Activision Blizzard's chief commercial officer, that they first hold an informal meeting of their tech leaders to brainstorm. "No biz folks, no me present, so that companies don't feel they need to send overly senior folk," Sweeney told him. Zerza liked the idea, replying, "We strongly believe that scale at the front end will benefit all of us and our players, plus increase the odds of success."

As they sorted the logistics of holding a summit, Sweeney probed Zerza, who had joined the video game company four years earlier from a career at Procter & Gamble, on his thinking. Sweeney shared some skepticism he held that a jointly operated store could unseat a dominant competitor, making a soft pitch for Epic's own app store ambitions. "We've envisioned launching the Epic Games store on Android and iOS with an 88/12 revenue split for all, and paying/making revenue guarantees to secure some key exclusive or co-exclusive titles," Sweeney shared in late June.

Zerza told him he was thinking of creating a consortium between his company, Epic, and Supercell with the goal of becoming the "Steam of Mobile," a single mobile game app store with a single payment system. Again, Sweeney remained skeptical. "Wouldn't realpolitik oriented participants . . . just take a consortium, or the threat of it, to Apple and Google to seek sweet deals just for themselves in exchange for abandoning it, while Apple and Google continue to tax everyone else?" His worries were prescient.

In fact, Zerza was contemplating the idea of a store; at least company records suggested as much. The strategy, too, received a code name: Project Boston. But it would become unclear how serious of an effort it really was and how much of the work was aimed at gaining leverage with Google—as Sweeney suspected. Internal company documents show Activision Blizzard was pursuing two paths: negotiating a deal with Google (then eventually Apple) and developing its own direct-to-consumer model through either an Activision Blizzard app store or something of a joint effort with rivals. The company wanted to develop a minimum viable product by the end of the year or early 2020 to show what was possible with its King mobile games—the popular *Candy Crush*—followed by something more robust later in 2020 and scaling it into 2021. Mock-ups of the app store looked similar to the feel of the Samsung store.

What it called "Path 1" with Google aimed to secure $100 million or more in value creation, roughly equivalent to what the company calculated would be reducing Google's 30 percent commission to 20 percent. And that seemed like the preferred path internal records noted, the efforts to create a store were "to put pressure on Google." It was what Sweeney had feared: Activision Blizzard was using its idea of a store as leverage in talks with Google. And that dynamic underscored why it was so hard for developers to rise up against Apple and Google. United, they might have a chance. Alone, however, very few had the ability to do much, as Sweeney was about to learn.

12
REVENGE OF GOOGLE

2019

The blaring headline on the slide to Jamie Rosenberg's team said it all: "Existential Question . . . How do we continue to keep Play as the preeminent distribution platform for Android?"

Tim Sweeney had found a way to exploit weaknesses in Google's digital realm that didn't exist in Apple's own digital empire. It became the Google executive Rosenberg's job to figure out a way to patch things up—something easier said than done. The fundamental issue was that Google didn't control all of its digital empire in the way Apple did. Whereas Apple could simply dictate rules because it owned both the software and the hardware in its iPhone empire, Google relied on phone makers adopting its Android software and then going along with its rules. Put another way, Google offered up a software world for the phone maker's hardware world—the two sides had to operate like a marriage. By early 2019, that was proving messy.

According to Google's own calculations, there were 2.3 billion Android devices in use that came preinstalled with its own apps—Maps, YouTube, and, perhaps most important, the Play store. And about 1 billion new devices coming online each year with those apps. Because Google gave away Android, the base layer of software for phones, to device makers, it relied on a web of contracts to create a Walled Garden experience to compete with Apple. Dense contracts on top of dense contracts, almost like a Russian doll of contracts, built on one another, each one unlocking something to entice the phone maker to go along with Google's wishes. At the heart of it was Google's desire for phone users—whether the phones were made by Samsung or someone else—to have an experience similar to, if not better than, what would happen if they had purchased an iPhone. That included having high-quality apps that provided functions that make up modern life, such as text messaging and directional maps. In Google executives' minds, this was especially important if users were coming from an iPhone; if the Android device didn't have the services and functions to begin with, then they might be more likely to return to Apple. Ensuring a new user had a good app store was part of that thinking. It allowed for users to quickly download their favorite apps.

A key contract that sat on top of the others was the big carrot Google used to get its way: the revenue share agreements, or RSA in Google-legal talk. Billions of dollars paid to phone makers, their cut of money made from search ads. Within this scheme, a handset maker had stair-step requirements it needed to meet to earn more money. At the high end was the premier tier, where a phone maker wasn't permitted to install any app store on their device other than Google Play. This meant that for customers taking the phone straight out of the box, there was only one way to easily acquire apps: through Google.

Another way Google aimed to control the digital empire was through a contract that had a mouthful of words for a title, the Mobile Application Distribution Agreement, or MADA. What the MADA offered a phone maker was the ability to preload the go-to apps from the Google library, the sorts of functions iPhone users naturally had: directional maps; email; the sorts of apps phone makers mostly wanted to give their customers. This was a key agreement, too, because it allowed the phone maker to incorporate Google's required software tools to allow popular apps, such as Facebook, to operate properly. In exchange, the phone maker had to preinstall Google Play on the default home screen. In order to qualify for MADA, phone makers had to agree to another contract—the Android Compatibility Commitment, or ACC—which aimed to ensure phones being sold were using the same version of Android that Google was using, ensuring its apps would be compatible across the phones made by different companies. A phone maker had to sign the ACC to qualify for the MADA and had to sign the MADA to qualify for revenue share.[*]

Samsung's 2017 MADA contract required eleven Google apps be preinstalled, including Maps, YouTube, and Play. Samsung, which had its own nascent app ecosystem ambitions, had objected to the terms, complaining that it wasn't allowed to pick which Google apps it wanted preinstalled. Some, including the Play app, had to be positioned on the default home screen. Some of those apps, like Play again, could not be deleted. Still, Google mostly got its way, agreeing to pay Samsung $800 million in 2017 and $1.2 billion in 2018, according to internal documents.

In 2019, it was time to begin negotiating new RSA agreements, and after watching what had just occurred with Epic and Samsung, Rosenberg and other Google executives wanted to reinforce their digital realm. That February, Rosenberg convened a working group to get his squad ready for war.

The major working hypotheses and assumptions underpinning their work were that Samsung presently was the only phone maker with sufficient market share to plausibly build an app store outside of China to rival Play in key markets—the Korean company sold 71 percent of premium Android phones, those selling for $600 or more—in the top ten countries were Play operated. Still, Chinese phone makers were a concern. While they lacked sufficient market share today to be an immediate concern outside China, they posed a long-term threat.

Then there were the big game developers, such as Activision Blizzard. With their existing user bases and muscle to potentially drive new distribution methods, they, too, could quickly chip away at Google's app store offerings by defecting.

To combat their concerns, they were plotting two different plans

---

[*] Except for Huawei, which Google was prohibited from entering contracts with because of a 2019 Department of Commerce ruling. Google's app store wasn't permitted in China.

of attack that they code-named Project Banyan and Project Hug. The first one aimed to keep Samsung and other phone makers from launching competing stores. The latter effort aimed to keep game developers from launching their own stores and keeping their content in the Google app store.

The effort was getting manpower and money. Rosenberg and his colleague Sameer Samat, were organizing a team of members across various functions—strategy, business operations, analytics, finance, business development, product, and engineering. A major bureaucratic undertaking that spoke to how seriously Google was taking the perceived threat of Epic's and Samsung's actions. Steering teams would meet roughly every two weeks for progress reviews, and Rosenberg and Samat would conduct milestone reviews every four weeks. No one should have thought that Rosenberg wasn't going to be deeply involved along the way, though. He and Samat were already making plans to aggressively engage with Samsung, which seemed like the biggest threat. Samsung devices were responsible for more than half of the revenue that Google's Play Store generated in 2018.

Within a few weeks, Rosenberg, Samat, and Hiroshi Lockheimer, a senior vice president, were on an airplane to South Korea to meet face-to-face with Samsung to present their proposal. The short of it was that Google offered to pay Samsung $200 million over four years in exchange for its app store to be subsumed within Play's billing system. The Play app would have a storefront within it that said "Samsung." Google would keep all revenue from in-app purchases made through Samsung's storefront. The idea, in general, wasn't a new one between the two sides. In 2015, Google had proposed something similar, but a deal didn't happen. Now Samsung seemed interested in talking more. A few weeks later, DJ Koh, the chief executive of Samsung's mobile business, spoke with Lockheimer by phone. "He assured me ('I give you my word') they understand now is a time to be working even closer together," Lockheimer told his colleagues afterward. Koh promised that his side would return with thoughts on Google's store issue. "Fingers crossed," Lockheimer concluded, "although I'm sure they will negotiate hard."

A day later, Rosenberg heard from Patrick Chomet, another senior

Samsung executive. "So you're basically asking us to get out of the store business," Chomet asked, according to Rosenberg's recounting. "I said, 'We're proposing that we focus together on helping you achieve your goals in a different way, and in a way that keeps us aligned to collaborate technically and with our message to the ecosystem.'"

"I would just encourage you to be as aggressive as possible on this in your proposal," Chomet responded. Rosenberg politely reminded him that Google had already made an offer. The ball was in Samsung's court. "Yes, yes," Chomet said. "I know it's with us, and we'll respond." As Rosenberg read his counterpart, it seemed clear Samsung wanted more than what Google had offered—much more.

*   *   *

Even if Jamie Rosenberg could negotiate deals to keep Samsung and other phone makers from siphoning off game developers with exclusive deals, there was the real threat that the game companies would try to make their own stores or, perhaps more threatening, band together to create a store for several different game makers. And there was the real possibility that Rosenberg might fail to persuade Samsung to go along with Google's wishes.

That's why Project Hug was launched. The job of overseeing it was given to Lawrence Koh, who had recently joined Google as the global head of game business development. He quickly learned that the company was scrambling to quell a growing rebellion among game developers who had been emboldened by Epic's actions.

The apps of just twenty-two developers accounted for 31 percent of the money spent in the Play Store. These twenty-two developers, which the team dubbed "beacons of the ecosystem," were unhappy, having expressed to Google their discontent with the revenue share model. Furthermore, Koh's team worried these developers would forgo launching upcoming major titles on Play, and since these companies were large enough and established enough, they could probably come up with the infrastructure required to launch their own stores like Epic. Google feared contagion, the idea of one gaming company trying it on its own spreading like a virus from one developer to the next until the

Play store emptied out. "We had concerns that once top developers took their gaming content off of Google Play that other developers would potentially follow suit," Lawrence Koh, Google's global head of gaming, would later say.

There was reason to worry. In recent months, executives at Tencent's Riot—maker of *League of Legends*—had disclosed that they were deciding whether to launch one of its new mega titles without Play. Activision Blizzard also had communicated that it was looking at starting its own competing Android store. "Some developers had expressed interest in coming up with their own technology solutions or coming out with a distribution mechanism to go direct to consumer on Android," Koh would later recall.

That spring, Koh and his team of roughly a dozen responsible for managing Google's relationship with video game developers prepared a presentation for the top echelon of Google leadership, the Business Council, which approved large spending. Project Hug needed money. Lots of it. The team's solution for countering Epic copycats was going to be costly.

To the Business Council, Koh's team warned that Sweeney's gambit made it easier for others to follow his lead. Everyone that followed, they felt, would have an easier time and a larger audience. In a big red box, the team prepared a slide that spelled out the danger: "The loss of top developers, either to competitors or by 'going-it-alone' on Android, would significantly impact Play's business." As they calculated it, developers, such as Epic, could afford to make less money outside Play because they wouldn't be paying Google's 30 percent fee. By the Project Hug team's math, more than *$2 billion in revenue*—maybe even as much as $6 billion—was at risk of being lost in the next three years alone. On top of that, they feared that as their own digital realm began showing cracks, consumers would become confused, creating an opening for Apple to exploit.

To counter that, the price tag for Project Hug was even more eye watering than what was proposed for Epic just months earlier. Project Hug would require $575 million and fifty-nine employees assigned to it—an audacious plan to run over three years to stanch the uprising. In exchange for payments in the forms of various credits, Google would demand developers agree to ship new games on Play at the same time as other platforms. It was dubbed a "sim ship" requirement, aimed at preventing the kinds of exclusive deals that Epic or Samsung might offer to be part of its store. It also ensured a developer couldn't bypass Play altogether.

Even before officially launching Project Hug, Google had been experimenting with the tactic, opening talks with King, part of Activision Blizzard, whose chief executive, Bobby Kotick, had become increasingly vocal against Google's 30 percent. By summer, the two sides had reached an agreement: King would get $20 million in ad credits from Google in exchange for sim shipping its games on Play. What would be offered later to Activision Blizzard was even more generous.

*   *   *

As Google's negotiations with Samsung dragged on that summer, Jamie Rosenberg, a senior Google executive, had a call with his counterpart at the Korean phone maker, Patrick Chomet, so they could talk about Samsung's position. Afterward, Rosenberg updated the Project Banyan team that Samsung was unhappy about Google's store proposal—in particular, about being required to use its in-app payment system. Samsung worried it didn't give the phone maker a way to build a direct relationship with its users or to build out its own payments system for the long term. In particular, Samsung saw the ability to have its own in-app payment system as the best way to establish a relationship with its users. The two men traded some ideas that didn't seem to go anywhere. Samsung suggested they could mitigate their two sides from competing with each other over app stores with some sort of "governance model," but Rosenberg dismissed that as overly difficult. Chomet returned to an idea that had been percolating throughout the talks: an app store revenue share, saying it would be a stronger proposal for them to consider. It was a topic that Google leadership had been pushing off. They agreed to keep talking.

The potential of not persuading Samsung to give up its store was creeping into Google executives' minds—even as they prepared another formal proposal ahead of flying back to Asia. If they failed, Samat, Rosenberg's colleague, told his colleagues, he figured Samsung's next steps would be pretty straightforward: They'd seek exclusive and unique content from game developers by taking less in the revenue share—perhaps 20 percent instead of 30 percent.

Rosenberg worried about the public relations battle that would ensue if Samsung made a big push to capture developers. "Despite our best efforts to secure distribution for our app store, if Samsung wins the hearts & minds of developers on this, it could create enormous pressure on us to unblock their opportunity one way or another," he continued. "I think devs will tolerate some premium pricing for distribution through Play/Google and all that we provide . . . but not a gap that wide." That said, he added, "I'm sure there is lots of legal advice on dos and don'ts here . . . feels like a risk that's very difficult to mitigate if we don't find a path to partnership."

Still, Rosenberg knew they had a backup plan in play—Project Hug—aimed at countering threats like Samsung. His team was also working feverishly to make sure smaller phone makers were neutralized. "I fully believe that on the basis of execution and momentum alone, we are well positioned to compete," Rosenberg told his colleagues. "We also likely have a healthy appetite to spend at scale to protect and grow the business."

Indeed, the team was worried not just about Samsung but also about the growing strength of Chinese phone makers, such as Xiaomi, Oppo, and Vivo, that were also investing in creating their own app and services ecosystem that could siphon away Google's business. Samsung-made phones constituted 26 percent of the market, while Chinese phone makers together had around 22 percent of the market (outside China).

To respond, Google was preparing a new round of contracts, dubbed RSA 3.0, to protect its digital empire. In a presentation to the Business Council, executives laid out a plan to spend roughly $15 billion during the next few years on phone makers other than Samsung. The spending would begin with $2.9 billion in 2020 and grow to $4.5 billion in 2023. Under the top-tier program, Google would share 20 percent of revenue made from its search business plus 16 percent of Play store revenue with the key Chinese phone makers for agreeing to forgo their alternative app stores and only preinstall Play. Smaller phone makers would get less money. (On top of that, Google planned to pay a total of $80 million to key Chinese phone makers to hedge against the potential loss of $1 billion in Play sales.)

It was a deal that most ended up agreeing to, further roiling Epic's ambitions.

*   *   *

Armed with a massive war chest, the Project Hug team began trying to lock down the biggest and most important game makers. None seemed bigger than Activision Blizzard, whose executives came prepared with their own big asks. Ahead of Thanksgiving, Armin Zerza, Activision Blizzard's savvy chief commercial officer (who was secretly talking with Tim Sweeney about creating an app store), told Google he wanted a deal that essentially lowered the Play store's take to 20 percent from 30 percent, or a difference of about $100 million a year.

Karen Beatty, the Google director heading up negotiations with Zerza, relayed his requests to the Project Hug team. The two sides were about $28 million apart from what Google had envisioned. Beatty had offered a deal that involved $72 million worth of credits. But more than that, Zerza made clear that Activision Blizzard wanted Google to buy exclusive rights to its games for esports broadcast on YouTube. What Zerza wanted was eye watering. "When I asked him how much it would cost, he asked if I was 'sitting down,'" she told her colleagues. Indeed, Zerza initially floated a demand of $70 million annually. "I responded that I would relay this information back to the teams, but that it was unlikely we would pay that much," she continued.

A few days later, Beatty and Zerza talked again. This time, he was presenting a counteroffer. Activision Blizzard wanted a three-year deal paying $100 million annually in credits for services including advertising and cloud computing. The math was complicated by the fact that he wanted YouTube to pay for exclusive rights to esports content. The "fuzzy" math, as Beatty relayed it to her colleagues, involved a payment structure of $50 million in year one and $60 million in year two plus a third-year option for around $70 million.* But Zerza was counting the cost to Google as roughly $20 million in year one because he calculated YouTube could generate ad and sponsorship revenue worth $30 million from Activision Blizzard's content. Others might look at it as he was asking for $130 million in year one and so on.

At first blush, the team sounded okay with most of the demands. The credits would be earned by Activision Blizzard, so asking for more meant ultimately the gaming company would need to spend more to unlock them. But the esports request had some balking. Ryan Wyatt, head of gaming for YouTube, pushed back on Lawrence Koh's team over the terms, saying the amount was too rich and threatened to set a bad precedent with "a negative ripple effect throughout esports."

"I have to say, this commitment to three years for ABK's Esports is not something I'd ever do in isolation," Wyatt told the team in an email. On his mind, in particular, was an upcoming media rights contract with Riot Games, maker of the popular *League of Legends*. The gaming company was owned by Tencent, which also held a stake in Activision Blizzard. "Understand that Riot will have these expectations in their 2021 media deal renewal, and that this payment will get back to them," Wyatt warned. "They will be able to argue their property is more valuable, and it is, and it could have serious cost implications for us." Instead, Wyatt suggested offering a much smaller, nonexclusive deal and to make Google a sponsor of their leagues. In his mind, they could reduce the annual price to around $7 million to $10 million in line with the current payment structure for Riot's *League of Legends*.

Wyatt wasn't the only Google executive to fret about the terms. Purnima Kochikar, a director dealing with apps and games on Play, wondered aloud to the team whether they needed to talk with Zerza's boss, CEO Bobby Kotick, about how such a deal would put them at a disadvantage with Apple. It was a message they had already delivered to Tencent leadership, which had told them it was "discussing all this as a 'family.'"

---

*The total three-year deal, worth possibly $180 million, would have been twice as large as the $90 million, two-year deal Activision Blizzard landed with Amazon.com's Twitch in 2018.

Two weeks later, the two sides were almost there but still not in agreement over the esports licensing. To make the request more palatable, Zerza offered to guarantee $15 million annually in YouTube advertising and sponsorships, Beatty told the team. Essentially, it appeared he was aiming to get a big headline number that valued Activision Blizzard's media rights at a high level, setting a benchmark for future negotiations once the contracts came to an end. Zerza wasn't shy about letting Google know he was talking to others. He suggested Amazon's Twitch, a live-streaming service popular with gamers, was in the mix, pressuring Activision Blizzard for an esports licensing deal. Activision Blizzard and Google could remove the esports license from their overall deal, but Zerza had made it clear he would still seek $100 million annually to offset Google's 30 percent fee. "If this deal falls through, he claims that they will launch their own mobile distribution platform (partnering with another 'major mobile company'—presume Epic)," Beatty cautioned. That said, she believed Activision Blizzard wanted the Google deal. "I do think they really want to do this deal with Google (vs competition)," she concluded.

To make it happen, Project Hug needed even more money. Their budget for Activision Blizzard was $115 million; locking down Activision Blizzard would take $245 million more. A week later, Koh's Project Hug team was back before Google's Business Council, spelling out the risks of not giving in to Activision Blizzard's demands in three bullet points:

- Activision has told us that they will build their own mobile store.

- Tencent portfolio exploring other business models to take revenues off Play.

- Epic's side loading *Fortnite* launcher turned into Epic Mobile Games store.

The Business Council agreed. Zerza quietly signed the Google agreement on January 24, 2020. By then, talks between Epic and Activision Blizzard had grown quiet without explanation from Zerza.

When Epic's Tim Sweeney saw a trade industry report that YouTube had acquired Activision Blizzard's esports rights for $160 million, the CEO suspected something was amiss. He dashed off an angry email to Don Harrison, Google's top executive over Android, that was steaming with grievances—from claiming Google was behind efforts to push negative media stories about Epic and stopping just short of accusing Google of paying Activision Blizzard $160 million to kill a potential app store deal with Epic.

"I'm concerned about what value was actually exchanged here, given that: 1) $160,000,000 of cash and in-kind cloud services seem to far exceed the value of those esports event rights as informed by own esports dealings on a product that's larger than the combination of all of those Activision products," Sweeney fumed. "And 2) after Epic had told Google we intended to launch Fortnite outside of Google Play in 2018, we were invited to a Google discussion with Jamie Rosenberg and Hiroshi Lockheimer who pitched a similar overarching business deal contingent on Epic abandoning our plans to compete with Google Play."

In short, Sweeney wanted to know if Activision Blizzard stopped talking to his company because Google paid them off. "Regarding the allegations you made . . . I've looked into these things and don't believe they are supported," Harrison replied in a late March 2020 email.

*   *   *

Project Banyan came to an end soon after Samsung put its demands in writing that summer. In a document it drew up, Samsung outlined what it wanted to achieve and how to do it. "The goal," it wrote, was to "prevent unnecessary competition on store." Executives wanted a five-year deal that paid $250 million. But more than that, Samsung wanted a 50 percent share of revenue generated from the Samsung app store within the app store. A store within a store.

It seemed a new approach was required. "We are halting work on this and not proceeding with the related work streams involving our respective app stores," Jamie Rosenberg told the Project Banyan team a few weeks after Samsung's term sheet arrived. "We're working internally with Legal to reevaluate our options, and we'll be back in touch once we have clear guidance on the direction this deal should take."

Instead, Google began talking about a new revenue share agreement, similar to what other phone makers had agreed to—except with much bigger dollar signs. During the negotiations that year, Google executives learned that Samsung was unhappy with how its deal with Epic on the *Fortnite* exclusive worked out. One executive had shared he wanted to avoid another deal like that in the future, Google's Sameer Samat told his colleagues after a conversation with the Samsung leader. Maybe it was more corporate politics; maybe it was an effort to smooth things over with a business partner. "I asked why—he said it was not good for google/samsung," Samat wrote. "I didn't fully understand."

Purnima Kochikar, another Google executive, had a hunch why. "It was a bad deal for Samsung," he replied. "They spent a very significant % of their marketing budget in return for a 4 DAY exclusive to distribute the Epic launcher, which enables Epic to distribute other apps and games on those devices! Whoever negotiated the deal did not understand the basics of the app ecosystem, undermined Samsung's own store ambitions, and potentially exposed their users to security risks. It is fortunate that this effort was not very successful.:-)"

As 2019 wound down, it was becoming increasingly clear that Sweeney's Android gamble had failed, especially compared with *Fortnite*'s performance on other platforms. The initial launch on Android saw early adopters patiently go through the steps of downloading *Fortnite* directly onto their devices. But then very quickly the rate of users fell off, far below what Sweeney's team had expected. The problems were twofold. Users who went to Epic's website faced a complicated seventeen-step process, including dire warning screens from Google.

It had no choice but to offer *Fortnite* through Play if it wanted to reach Android customers. But again, Sweeney was defiant. Before submitting the *Fortnite* app code to the store, he wrote Google to say he was going to use Epic's own payment system rather than Google's. With Christmas coming, he argued that it was the perfect time to launch *Fortnite* on Google's app store. "Confidentially," he told Rosenberg, Epic was planning a partnership with Disney tied to the new

Star Wars: The Rise of Skywalker movie. In *Fortnite*, the movie's director, J. J. Abrams, was going to premiere a clip of the highly anticipated movie along with an in-game *Millennium Falcon* aerial battle followed by other tie-ins throughout the month. "We anticipate all-time records in player concurrency and stream viewership with these events," Sweeney wrote.

Sweeney urged Google to allow it to use its own billing system, noting it had already processed more than $1.2 billion of sales through its other venues. "The future of smartphone is in open platforms where developers can choose freely among storefronts, in-app payment services, cloud services, and engines," Sweeney reiterated. "This future is going to come about one way or another." And if his rhetoric wasn't already heated enough, he suggested that a bigger fight lay on the horizon if Google didn't submit. "Throughout the harsh public and private scrutiny that will be directed at commerce-locked digital ecosystems throughout 2020, I hope we can hold up Android and Google Play as a shining example of best practices that uphold the legal obligations of the two dominant companies in the $70 billion annual app market," he concluded.

It is hard not to think that Sweeney knew his demands would be immediately rejected. He would say otherwise, though. He argued that Google was inconsistent when enforcing its rules, an example, perhaps, of how behind the scenes the tech giant, unlike Apple, was willing to bend its position for certain players in exchange for going along to get along. Google had essentially offered the same things to Epic that it gave Activision Blizzard King and other game developers, but such a deal was not to Sweeney's principled liking. He didn't want to be tied into submission. He wanted freedom.

Not surprisingly, Google would ultimately reject the *Fortnite* submission because of the billing issue. This time around, however, Google prepared for a public relations battle, sensing Sweeney would go public again. As internal emails flew around Google, Samat saw Epic's actions against Google as hypocritical. "For over a year they have listed Fortnite in Apple's App Store, use their billing, and pay them 30%—but there [sic] current rhetoric and protest seems targeted only at Android," he sniffed. Hiroshi Lockheimer agreed as he warned senior leaders that they were expecting a "significant amount of press" around Epic's actions. As they worked on their public arguments against Epic, however, the language of hypocrisy was briefly dropped. Chief Financial Officer Ruth Porat disagreed with that decision. "Seems to me that it is very relevant that they pay appl for the benefit of their store—why the decision to delete that reference from our comms?" she asked in an email.

Kent Walker, Google's top lawyer, chimed in that it had been his call. "Apparently Epic argues that Apple's App Store works better for them and other developers than our Play Store does, and I wasn't sure that we wanted to open that issue," he responded.

The senior executives felt as if they were in a bind. They wanted to get their full side of the story out and not let Epic paint them in a negative light, but also realized that a big response by Google would generate even more attention. After receiving some feedback, Walker suggested they find a way to get their side of the story out. It was important to spotlight that Epic was refraining from criticizing Apple not because of concerns about differences in services between Android and Apple but "because they're afraid of retaliation by Apple (which has more market power ability to promote or not promote apps, etc.)," he wrote.

In short order, Sweeney's demands were leaked to a tech enthusiast website called 9to5Google, where the reporter Abner Li wrote a story with the headline EXCLUSIVE: EPIC SUBMITTING FORTNITE FOR ANDROID TO PLAY STORE IN HOPES OF SPECIAL BILLING EXCEPTION.

When Sweeney read that headline, he practically seethed: *special billing exception*. The mere suggestion that Sweeney was asking for *special* treatment would irk the man for years to come, ratcheting the temperature up between him and his Google counterparts, whom he accused of leaking his plans and twisting his words. Sweeney issued a public statement: "Epic doesn't seek a special exception for ourselves; rather we expect to see a general change to smartphone industry practices in this regard."

Privately, he plotted his next step. First, he would direct his team to comply with Google's rules, release *Fortnite* on Play, and give in to the 30 percent commission. But that wasn't the end of things for Sweeney. No, he was preparing a trap.

13

LIBERTY OR DEPLATFORM

2020

Video games are like TV shows. What's a blockbuster one season might not tickle everyone's fancy a year later. To the outside world, *Fortnite* had been the hottest thing around. But as 2019 played out, internally the numbers weren't as hot as Tim Sweeney and his team had expected, hindering the company's growth. Epic Games' sales fell in the second quarter from a year earlier by 27 percent and were on pace to generate $300 million less in the third quarter than they had planned.

Sweeney's gamble against Google wasn't helping. Before launching on Android outside the Google Play Store, Epic's numbers people had expected much higher downloads than they were seeing. The challenge of side loading, plus the bad PR about security in those early days, had all seemed to dampen those projections. And after Sweeney decided to reverse course, putting *Fortnite* on the Google app store in early 2020, Samsung executives expressed their unhappiness. The South Korean company's executives had stuck their necks out for Epic to help it go rogue, giving it a sweetheart deal to be on their own rival app store. Now Samsung was signaling that it wanted to renegotiate. In the first three months of 2020, revenue continued to fall. Sweeney was stuck between a rock and a rock; combined, Apple and Google represented 99 percent of the mobile market.

The bulk of Epic's success was on video game consoles—which had a limited audience. The biggest generator of *Fortnite* sales came from Sony's PlayStation, which was on track to have paid $6 billion since launching the game through 2020. Microsoft's Xbox was the next best, earning it $3.5 billion in that same time period followed by $1.1 billion from Nintendo's Switch. Since launching *Fortnite* on the Apple App Store in early 2018, Epic had earned roughly half as much as it received from Nintendo, or about $600 million. Or in other words, Apple was still just a small part of the overall *Fortnite* pie.

Sweeney had hoped bringing *Fortnite* to mobile devices would spark further interest among younger people in video game consoles, such as Xbox, while also expanding the pie altogether. For many, the reality was that iPhones and Galaxy phones were the new video game players. On top of that, these devices had become the gateway to a new digital world that was quickly changing. Developments in virtual reality and artificial intelligence held the promise of bringing science fiction to real life, a thought that Sweeney had increasingly become enthralled with as he imagined the potential. In many ways, *Fortnite* was a building block to that new realm that he, and others, called the "metaverse." He saw it as a place not just for gaming but for socializing, a vision for the future inspired by the book from almost a decade earlier, *Ready Player One*. A place for creators to entertain; a marketplace. Apple, among other tech giants, was busy working to develop its own virtual reality headsets that game makers like Epic would use for their digital worlds. But if Apple's 30 percent fee extended to that new world, Sweeney worried the economics wouldn't work, especially if *Fortnite* was becoming its own Super App, a place for commerce, communication, and entertainment that developers would create for him.

Just a couple of years earlier, Sweeney had savvily pushed the video game machine industry to accept interoperability, opening the door for *Fortnite* to create a digital world untied to a single hardware platform. Now he was clearly setting his sights on what he saw as the next impediment to ushering in a new generation of platforms—the money rails. And while Epic was about to take a public loss by making *Fortnite* available in the coming weeks on Google Play, he wasn't done fighting. He was going to make that much clear on one of the industry's biggest events: the Academy of Interactive Arts & Sciences' annual awards celebration.

At the video game industry's equivalent of the Oscars, Sweeney took the stage at the Aria Resort & Casino in Las Vegas to give a keynote speech that had been billed as "The Times They Are A-Changin'." While he didn't mention the term "metaverse," he laid out how he saw technology trends evolving that would remake the role of video games in modern life, an evolution he was already seeing with *Fortnite*, which he would say had become more than just a game and was a platform unto itself: part communications, part social network, part marketplace. Standing in the way of that vision were Apple and Google and their payment systems that collected an unfair tax. He noted his fight with Google and how the tech giant had refused *Fortnite* when submitted with an alternative payment system. "That's unfair, and that needs to change and it will change," Sweeney said. "And I think we as an industry need to use all of our resources to ensure that all these platforms are opened up to us as speedily as possible."

*　　*　　*

No sooner had *Fortnite* launched on Google than Sweeney and his top deputies were plotting their next move, a secret plan they would code-name Project Liberty. Among the tight circle, the outline of the plan was coming together that May. Epic was going to launch its own in-app payment store and highlight how it was cheaper than those required by Apple and Google. While they were narrowing in on how to pull off the scheme technically, they were still debating how to come out the hero in the public's mind—especially if *Fortnite* play was disrupted on mobile devices. "How do we not look like the bad guys and what is our strategy to counter Apple and Google's lobbying as I assume that will be in overdrive as they will treat this as an existential threat," Daniel Vogel, Epic's chief operating officer, asked his colleagues. Project Liberty would be more than an engineering feat. He suggested the company's legal and public relations teams be brought into the fold.

Sweeney agreed and told one of his executives he wanted to be added to all meetings on the topic. "Please keep me in the loop on this topic 100%," he instructed. To make things clear, as if they weren't already, Mark Rein, the longtime Sweeney deputy, stressed to the group to expect swift and strong responses to their attacks. While they might have seen themselves as storming the beaches of Normandy, others might see them as bombing Pearl Harbor. "We should . . . realize that there's a better than 50% chance Apple and Google will immediately remove the games from their stores the minute we do this," he told them. "They may also sue us to make an example."

*　　*　　*

A few weeks later, Matt Weissinger, Epic's head of marketing, was getting some outside advice on how to respond in a potential PR war with Apple and Google. It didn't sound great. The challenges were many, according to a presentation created by Lane Kasselman, a communication strategist from Greenbrier that Epic had paid $100,000 for help. Kasselman had a special understanding of guerrilla PR tactics and how Big Tech operated, having been one of the first hundred employees at Uber. As head of communications, he helped fight regulations that threatened to derail the ride-share company's business model. He'd gone on to co-found Greenbrier, a boutique public relations firm that was acquired in 2018 by a firm started by the former Obama White House deputy chief of staff Jim Messina.

"Epic is seen as a successful major company earning billions of dollars a year, and is not seen by press and players as an immediate sympathetic figure in a royalty share battle," Kasselman's presentation said. He pointed to Epic's experience in the past two years in its public feud with Google over the Play store as a test run for how things might play in a bigger war. Initial press reports of Epic bypassing Play was mixed, with "heavy debate on whether or not Epic can be successful on Android without the support of the Play Store," he concluded. By December 2019, press coverage had turned negative and cynical against Epic when the story was leaked that the game developer was seeking an exemption from the 30 percent fee: "They still don't want to play by the same rules that everyone else has to." Epic should expect more of that. A lot more.

Additionally, Kasselman suspected the media would be naturally inclined to default to the big tech companies' narratives, quickly losing interest in the story after the initial salvo until major decisions or developments occurred. Because of that, it was going to be hard to keep any momentum going, all while *Fortnite* could very well be missing in action from the two ecosystems. Not said but clearly part of the equation: Would *Fortnite* even be relevant once all of the fighting was over?

To build lasting momentum, Kasselman recommended Epic create a coalition of developers to turn the fight into a broader public debate to keep the issue top of mind. This was to be a fight not just about Epic but about an entire industry being mistreated. Such a group would take a couple of months to get started as well as $80,000 to $100,000 of seed funding to conduct opinion polls on whether app prices are too high and to run ads fomenting public support.

"Go the nuclear option," Weissinger wrote to himself in notes, presumably about Project Liberty. "It becomes the catalyst for the coalition, which is up and running, and can dimensionalize the battle for us."

Even so, there seemed to be a strong suggestion that Epic might want to avoid going after both Google and Apple at the same time. Weissinger's notes mentioned, "We recommend we go the nuclear option and submit onto Google Play" and, in particular, "Apple has no problem going down a litigation path, and they will not cave in a week or a month."

In theory, creating a coalition of like-minded developers was a great idea. Except, Tim Sweeney had essentially tried to do that a year earlier when he approached other game makers about creating a store. Through the process, Activision Blizzard used the threat of Epic as a bargaining chip in negotiations with Google for a better deal for its own business. It was an industry accustomed to bare-knuckle fights, everyone scrambling to get themselves a little better deal and everyone living in fear of Apple.

As Weissinger thought about building a coalition for Project Liberty, he jotted down some ideas of steps they needed to take: Create a narrative that we are benevolent. Spend some money, maybe $20,000, to test some messaging, get data to show that consumers will support their fight, write a one-page manifesto to recruit partners. But who might be Epic's partners? Two or three other companies would need to get on board as founding members of the coalition. Ideally, he wrote, it would be a company that's already a counterparty. Somebody in the fight.

Spotify.

*   *   *

The idea of going public against Apple carried a lot of risk. But what other option did Tim Sweeney have? Epic had already failed to pry open Google's digital realm. It was a fight that had turned ugly. The tech giant suggested Sweeney was okay with Apple's nearly identical approach since he wasn't fighting the iPhone maker. Of course, there wasn't a way to try to go around Apple's Walled Garden, no side-loading option to test out and fail on like with Android. Apple didn't even want to talk about ideas for change. By Sweeney's count, he and Rein had made a dozen attempts to engage Apple on the matter. There was his email to Apple's CEO, Tim Cook, in 2015 and his attempt in 2018 to meet with the marketing executive Greg Joswiak, but those efforts hadn't gone anywhere. There was but a chance five-minute encounter with Matt Fischer, an Apple executive who oversaw the App Store business. That, too, resulted in nothing.

Still, before going nuclear, Sweeney wanted to make every effort to go through the front door, even while plotting a sneak attack. On June 30, he wrote Cook; the App Store leader, Phil Schiller; and some other executives a lengthy email titled "Consumer Choice & Competition." In the note, he laid out why Epic wanted to offer a competing app store on the iPhone and competing payment system. "If Epic were allowed to provide these options to iOS device users, consumers would have an opportunity to pay less for digital products and developers would earn more from their sales," Sweeney wrote. He added that he hoped Apple would make such options available to all developers.

Then the kicker: He wanted Apple to agree in principle within two weeks. "If we do not receive your confirmation, we will understand that Apple is not willing to make the changes necessary to allow us to provide Android customers with the option of choosing their app store and payment processing system," Sweeney concluded.

Yes, he mistakenly wrote "Android" rather than "Apple"—an embarrassing typo that tipped his hand that Epic was also writing a similar demand to Google.

Sweeney's prelude to war came via a form letter.

*   *   *

July began with a meeting at Epic to go over the status of Project Liberty. If Apple and Google rejected their entry, then they planned to move forward with their attack. As they looked at the calendar, they were targeting mid-August, in particular August 13—which would fall after a planned August 4 update of the game but before the August 27 start of a new season of *Fortnite*, when users would be clamoring to play the new version. The new season was bound to generate a lot of attention given it had a Marvel superhero character tie-in.

If Apple and Google removed *Fortnite* from their stores, as Epic expected, mobile users who didn't already have the game on their phones wouldn't be able to install it. Those who did have the game could continue playing but not upgrade to the latest version. However, it wouldn't be long before those players with dated software would be cut off from the broader *Fortnite* universe, unable to compete with players on other devices with the newest programs. Apple and Google Play users were left behind when a new season of *Fortnite* began, left only to play among their own ecosystems.

A key part of the plan was launching a price cut for users of Epic's own payment system. What Tim Sweeney wanted to do was show the world that Apple's business model was misunderstood. It wasn't about software distribution, in his opinion; it was about exercising total control of its software and using that control to deny access to third-party software.

Epic's engineers were busy. About two hundred people were now assigned to Project Liberty, including specialists to try to reverse engineer the hotfix code to see if it could be detected. More simply put, they were going to get one of their hackers to hack the hack to see if it could be independently discovered by Apple. As long as it wasn't discovered, they were set.

The self-imposed deadline was fast approaching when Apple on July 10 declined Sweeney's demands, defending its store. Neither Tim Cook nor Phil Schiller responded. Instead, the reaction came from Apple's legal office, which noted that if Epic didn't like its terms, it had other ways of selling his game while also pointing out that the game maker collected royalties of its own from games built on its Unreal Engine. "Yet somehow, you believe Apple has no right to do the same, and want all the benefits Apple and the App Store provide without having to pay a penny," Apple concluded. "Apple cannot bow to that unreasonable demand."

Sweeney took issue with Apple's responding with a letter from a lawyer. "It's a sad state of affairs that Apple's senior executives would hand Epic's sincere request off to Apple's legal team to respond with such a self-righteous and self-serving screed—only lawyers could pretend that Apple is protecting consumers by denying choice in payments and stores to owners of iOS devices," Sweeney replied in a new message to Cook and Schiller. He concluded that Epic "is in a state of substantial disagreement with Apple's policy and practices, and we will continue to pursue this, as we have done in the past to address other injustices in our industry."

*   *   *

A growing political backlash against Big Tech was becoming palatable in Washington, D.C., where for years politicians seemed to take pride in the nation's biggest companies. Now both sides of the political aisle questioned their reach. For Democrats, these companies represented how big business was hindering competition. For Republicans, they worried about how the likes of Facebook and Google were limiting speech. The Trump administration's Justice Department in 2019 had opened broad inquiries into Big Tech, including Apple, to see if they were stifling competition. And the subject was gaining attention on the Hill that summer. As the House Judiciary Committee's antitrust subcommittee prepared to hold a hearing on Big Tech, Apple representatives felt unfairly grouped with Amazon.com, Facebook, and Google, which had long faced real questions about how they operated. In response, they were told Apple was the largest company in the United States by market value and couldn't be excluded from the conversation—even if not everyone yet fully understood why their App Store might be problematic. The CEOs of the four companies were called before the committee—remotely because of COVID-19 restrictions—for more than four hours of questions, a scene reminiscent of when Big Tobacco was called before Congress to answer for its sins.

"As gatekeepers to the digital economy, these platforms enjoy the power to pick winners and losers, shake down small businesses and enrich themselves while choking off competitors," Representative David Cicilline, a Democrat from Rhode Island and chairman of the House Judiciary Committee's antitrust subcommittee, began the hearing. "Our founders would not bow before a king. Nor should we bow before the emperors of the online economy."

When it came time for Cook to speak, he was deferential but firm. "I'm here today because scrutiny is appropriate," Cook said. "We approach this process with respect and humility, but we make no concession on the facts. What began as 500 apps is now more than 1.7 million, only 60 of which are Apple software. If Apple is a gatekeeper, what we've done is open the gate wider. We want to get every app we can on the store, not keep them off."

As NPR would note after the fact, Apple received less attention throughout the hearing. When it did, that attention was diffused by what Apple was supposedly doing wrong. Representative Val Demings, a Democrat from Florida, questioned Cook on removing third-party parental control apps. The House Judiciary Committee chair, Jerrold Nadler, a Democrat from New York, pushed Cook on media reports that Apple had demanded Airbnb and ClassPass share revenue for new virtual experiences launched during the pandemic. "Isn't this pandemic profiteering?" the congressman asked.

"The pandemic is a tragedy, and it's hurting Americans and many people from all around the world and we would never take advantage of that," Cook said. "I believe the cases that you're talking about are cases where something has moved to a digital service, which technically does need to go through our commission model. But in both of the cases I am aware of, we are working with the developers."

Representative Hank Johnson, a Democrat from Georgia, narrowed in on the power Apple held over the App Store and businesses trying to do business on it. The congressman suggested Apple treated some companies differently, such as Amazon and Baidu. "Has Apple ever retaliated against or disadvantaged a developer who went public about their frustrations with the App Store?" he asked.

"Sir," Cook replied, "we do not retaliate or bully people. It's strongly against our company culture."

*   *   *

Tim Cook's appearance before the House committee came as Apple's business was never better—iPhone and computer sales soared in 2020 as people stuck at home during the pandemic turned to digital devices to occupy their time. His gamble on the digital services was also looking especially smart, fueling record growth in the quarter that was announced at the end of July.

Still, the year 2020 was a time of reflection for a lot of people, including Phil Schiller, who began to ask if it was time to make a change. At sixty years old, Schiller had spent much of his adult life working for Apple—about twenty years in the same role as head of marketing. Work-life balance among Apple senior executives was easy. Work was their lives. And Schiller was no different, working eighty hours a week. He hadn't taken a week of vacation in, maybe, a decade. He didn't want to step away from Apple but began thinking how to step back. On August 4, Apple announced Schiller would give up the reins on Apple's marketing for a new role as "Apple Fellow." That was a title that had been handed out on rare occasions before, mostly to those stepping far back from the day-to-day, such as co-founder Steve Wozniak. And the news was largely interpreted by outsiders as a sign

Schiller was heading toward retirement via a soft landing after years of service rendered. Maybe it was. For a hard-charging guy like Schiller, the thought of working sixty hours . . . or forty hours a week may have seemed like coasting. In reality, Schiller was about to be thrust into the heart of battle. There would be no time for quiet quitting. He would be asked to become the face defending his life's work: the App Store.

*   *   *

COVID-19 was turning out to be good for business for Epic, too. With everyone stuck at home because of the pandemic, many people were turning to video games as a diversion.

The uptick in business helped Epic turn the tide from the early few months of the year when sales in the first quarter fell almost 40 percent from the same period a year earlier. Now Epic was humming along, outperforming its expectations by 70 percent, generating $2 billion in the first six months of the year, and projecting continued strong sales to finish the year with almost $5 billion in 2020 sales, or 17 percent better than in 2019. On an adjusted earnings basis, it expected a 24 percent gain for the year compared with the previous twelve months.

Tim Sweeney's team came armed with those numbers to a board of directors update on the status of Project Liberty in late July. Sweeney was setting the table for why it was time to strike Apple and Google. "The time is Now," the presentation said. Essentially, he was betting that the bounce back in sales during COVID could mostly offset any lost revenue from taking on Apple and Google in 2020. The risks were very real, including loss of sales. Sweeney and his team expected a big fight. "Possible negative player sentiment as Apple and Google PR campaigns direct blame to Epic," the board was warned. But the bottom line: "Epic is prepared to fight as long as it takes and will not back down from direct payments."

Sweeney was looking to a future in which *Fortnite* would evolve into its own platform where creators would build content in what he called the "metaverse," a virtual reality space where games, entertainment, and work could take place. The idea of the metaverse was becoming almost universally embraced by certain tech leaders.

Facebook's co-founder Mark Zuckerberg became so excited about the potential that he renamed his company Meta Platforms as he pursued a VR platform that he gambled could finally give him hardware product to displace the iPhone. How that world would exist and who would control it was at the core of Sweeney's fight. "Fortnite as a competitive platform for user content requires sharing a majority of profit with creators," the board was told. The idea of 30 percent of the revenue going to Apple and Google would threaten that vision. Sweeney wanted to "solve this problem before [augmented reality] takes off and that rate is set at 30%."

*   *   *

At the core of Project Liberty was the surprise attack. It would begin almost like something out of *Mission: Impossible*, but instead of Tim Sweeney rappelling down from a ceiling, Epic was going to put the equivalent of a Trojan horse inside the *Fortnite* software that Epic would upload to the Apple and Google app stores for users to download. The app update included a hidden payment system outside the app stores.

The scheme depended on Epic's servers to eventually talk to the software downloaded onto devices, giving it new instructions on what and when to display the hidden feature. So-called hot fixes were commonly used to make tweaks to a game on a day-to-day basis. In the normal course of business, Epic might use a hotfix to allow players access to a specific type of game for a period of time or new cosmetic for their avatar or tweak the competition in the game to make a weapon less effective. Hotfixes might happen on a weekly basis. But what Epic was planning was no minor tweak. It was a very major change.

In early August, Epic submitted updates of *Fortnite* to Apple that included its special Project Liberty hot fix that sat dormant for two weeks until the time was right to flip the switch.

Andrew Grant, a technical director in engineering, was assigned to oversee the hotfix and contingency planning if Apple responded by blocking updates of *Fortnite*. When submitting the update to Apple, they excluded mention of the hotfix in the developers' notes that the reviewers would have relied upon. Apple's contract required Epic to disclose fundamental changes. "I knew we were doing something that Apple would be unhappy with," he later recalled.

Another key part of the plan was coming into place that August: the coalition of the like-minded app developers. Spotify had been an obvious first choice. Daniel Ek, Spotify's CEO, had been one of the rare tech executives willing to stick his neck out against the iPhone maker. But Spotify's top lawyer, Horacio Gutierrez, who had redirected the company's fight against Apple to Europe from Washington, wasn't interested in joining an astroturf organization that was only about drumming up support for Epic's legal battle. If such a group was going to exist, it needed to have a broader remit. He wasn't just advocating for change in Brussels; he was pushing regulators in the U.K., Japan, and elsewhere. Soon the group, the Coalition for App Fairness, would be representing Epic and Spotify in fights around the world. Their efforts were joined by another big name, the Match Group, the online dating company whose stable of popular apps included Tinder, an app known by users for swiping on potential partners' photos to signal interest.

As the time of Epic's attack grew closer, Sweeney began reaching out to allies at Sony, Microsoft, and Nintendo, alerting them to upcoming price changes in *Fortnite*. He told an ally at Microsoft on the eve of launching the attack, "You'll enjoy the upcoming fireworks show."

Then, at 2:00 a.m. at Apple's headquarters in Cupertino, Tim Cook and Phil Schiller received an email from Sweeney. This time a declaration of war. "I'm writing to tell you that Epic will no longer adhere to Apple's payment processing restrictions," Sweeney began before detailing Epic's payment system. "We choose to follow this path in firm belief that history and law are on our side. Smartphones are essential computing devices that people use to live their lives and conduct their business. Apple's position that its manufacture of a device gives it free rein to control, restrict, and tax commerce by consumers and creative expression by developers is repugnant to the principles of a free society."

He followed up with Epic flipping the switch on the hotfix, revealing its own in-app payment system for Apple and Google users. At the same time, it also lowered its in-app prices by about 20 percent, and launched a marketing campaign it dubbed "Fortnite Mega Drop." In theory, Apple and Google users in Epic's payment system could enjoy the discount but not if they used the tech companies' in-app system. By the end of the day, both Apple and Google booted *Fortnite* from their stores. The second part of Sweeney's plan was triggered: a court battle. Or put another way, Epic was now in full-blown war with the world's most powerful companies. To mark the occasion, Sweeney released a video on social media of his *Fortnite* characters in a reimagined version of Apple's famous "1984" commercial aired years earlier to mock IBM. In Epic's version, the Big Brother character on a giant screen was replaced with a talking apple, which happened to have a bite missing from its head like Apple's logo and a worm sticking out. "For years, they have given us their sons, their labor, and their dreams. In exchange, we have taken our tribute, our profits, our control." After a *Fortnite* character breaks the screen, the spot ends with a call to arms. "Epic Games has defied the App Store Monopoly. In retaliation, Apple is blocking Fortnite from a billion devices. Join the fight to stop 2020 from becoming '1984.'"

WILD WEST OF LAW

2020

Apple took the bait and then some. It not only booted *Fortnite* from its App Store but also delivered its own threat to go nuclear against Epic Games over its insubordination: It gave the game maker just fourteen days to comply with its rules or else it would block it from software tools needed for its Unreal Engine to run on iOS. That, in turn, would disrupt an entire ecosystem of game developers who relied on Epic for its world-building programs.

Epic's high-powered legal team, which included Christine Varney, the former Justice Department antitrust chief whose office had gone after Apple years earlier, had been ready for the first move. It filed lawsuits against Apple and Google the day *Fortnite* was kicked out of the apps stores. The totality of the filing underscored how long Epic had been working to spring the trap. The arguments ran sixty-two pages.

The day after Epic filed its lawsuit, Apple sent the game maker another letter, giving it fourteen days. Epic lawyers raced to court for a more urgent restraining order to prevent Apple's move, which they called retaliatory and potentially "catastrophic."

"Not content simply to remove Fortnite from the App Store, Apple is attacking Epic's entire business in unrelated areas," the filing began.

The case was assigned to the U.S. district judge Yvonne Gonzalez Rogers. From her courtroom in Oakland, Rogers was familiar with both companies, having had cases involving both of them appear before her. Just weeks earlier, she had Epic in her courtroom on another matter. "Is 'Fortnite' as big a deal as they're saying it is?" the judge asked the lawyers.

"My daughters think it's kind of cool that I get to work for them," Epic's lawyer responded. Gonzalez Rogers offered her own son as a rebuttal.

"He wanted to play videogames," the fifty-six-year-old judge said. "I told him 'no' so many times that out of paper, he created his own and he made little cards that he would put into the sleeve so he could play games. He's now an aeronautical engineer, and I am quite happy that I said 'no.'"[1]

She approached her work with a certain no-nonsense manner. As an undergraduate, she earned money by cleaning houses and cutting grass to help pay for her tuition at Princeton University. She received her law degree from the University of Texas at Austin, then had a career as a litigator in San Francisco until she was appointed to the bench on the Alameda County Superior Court. In 2011, President Barack Obama elevated her to the federal bench. Her early cases involving Apple would foreshadow just how hard of a fight Epic had ahead of it. Now Gonzalez Rogers was overseeing another Apple battle, this time under COVID protocols. Instead of routine courtroom hearings, the initial proceedings were being held via Zoom video conferences. Still, her signature blunt approach was on display. "Both Apple and Epic Games are in my courtroom all the time—all the time," she said as she interrupted Epic's lawyer during a hearing on the Wednesday after the Project Liberty attack was launched. "Obviously, there's lots of news articles happening about what's going on and the alleged harm and everything else."

"The underlying issues in this case are going to be decided in a reasoned, regular process," she continued. At the heart of things, she wanted to know if Apple's threat against the Unreal Engine could be rolled back to the point where third parties weren't harmed while the dispute between Epic and Apple was properly litigated—a prospect that in normal circumstances might take years. "You are all pre-eminent litigators," she concluded. "And the question is can you pick up the phone and figure out how to resolve this in a way so that the underlying litigation can proceed in a normal way?"

Apple's outside lawyer, Richard Doren, responded that the issue at hand was much more than about money but about the attack on Apple's reputation. "We do agree that we can go back to the status quo until these issues are resolved," he told her. "But it would need to be the status quo before this hotfix was implemented by Epic in the App Store."

To Epic, the threats against its Unreal Engine were unrelated to the hotfix attack. "That has resulted immediately in uncertainty from the developer community as to whether or not they actually have a plat-form they can develop on which includes not only new applications but also updates and fixes and things of that nature," Katherine Forrest, Epic's outside lawyer, told the judge. In short, Epic wanted the judge to let the developer community know they were safe while the two sides duked it out in court on the payment issue.

"The reason that risks are presented to Epic that were just framed is because Epic actually changed the status quo," Doren responded. "And is now seeking to freeze things in this new . . . world that it created."

As the two sides continued to go back and forth, the judge asked if she would need to deal with the question by Monday. "What you want me to do is—you billion-dollar companies—want me to drop everything and deal with your papers over the weekend because you can't on your own seem to put a temporary stop so we can deal with this," she asked.

Apple conceded it could wait for the judge to hear the matter at her convenience, noting its deadline to Epic was still days away on Friday, August 28. Predictably, Epic urged the judge to address the matter as quickly as possible. The threat was not just about August 28 in Epic's opinion but about the uncertainty that had been created. "Look, Epic Games knew what it was doing, right?" the judge interjected. "It de-cided to take a stand, which is fine. It implemented what they're calling this 'hotfix' and immediately filed a complaint. So it knew exactly what it was doing.

"The problem is that it didn't quite anticipate the kind of response that it got from Apple. You created the situation by doing what they did. You don't come into this fight with clean hands." That said, she agreed to a hearing on the following Monday.

*   *   *

To prepare for Monday's make-or-break hearing, both sides filed lengthy responses. That included a statement from Phil Schiller, head of the App Store, who previewed Apple's general defense for months and years to come: Apple and Epic had a contract—a contract that Sweeney's company had willingly signed and benefited from for years. And when Apple didn't bow to Sweeney's demands that it change its long-standing business practices at a whim, Epic responded by illegally breaking that contract. "When Apple refused to fundamentally alter the way it does business to appease Epic, Epic resorted to sudden, unilateral action that blatantly breached its contracts with Apple, and simultaneously filed this lawsuit, which seeks to justify its deliberate breaches after the fact," Schiller said. "Epic essentially granted itself preferential treatment vis-à-vis all other developers who offer in-app purchases on Apple's iOS platform, including Epic's direct competitors," he continued.

Apple, Schiller noted, had offered to allow *Fortnite* back into the App Store if only Epic removed its offending payment system. He argued iPhone maker's threats against Unreal Engine were in line with how it dealt with other bad-faith developers. "As in the past, where Apple has terminated a developer account for bad faith or deceptive conduct like what Epic did here, we have also terminated accounts that we know to be affiliated with the offending account," he told the court.

Sweeney's move was about more than Apple losing out on fees. Schiller warned the judge that Epic's actions threatened the very core of what the company had built with the iPhone and the App Store, which he said was more than just a marketplace but a large number of tools, technologies, and services offered to developers. "This entire ecosystem would be in jeopardy if developers are allowed to breach their agreement without consequences as Epic has done," he told the court. "If every developer is free to breach its contracts with Apple . . . then Apple's App Store cannot deliver the many benefits to consumers and developers that it currently does."

During the Monday hearing, Apple's lawyer worked to highlight to the judge the underhanded nature of the hotfix scheme. "Apple's business model and its most fundamental promise to its customers is that they will be . . . safe, their privacy will be secured, and there will not be malware," the lawyer told the judge. "There will not be unknown codes migrating into the App Store. And what Epic has done is to break that model . . . and profit by it, and place the customers in the middle."

In response, Epic tried to minimize the suggestions of deceit. "Hotfixes, which are being described as something nefarious and something novel, are neither nefarious nor novel," Katherine Forrest told the judge. "They happen all the time."

"Epic Games didn't tell Apple that you had code in there which would allow you to collect directly from your consumers, right?" the judge interrupted.

"Your Honor," Forrest replied, "we didn't hide it. We never ever go through every single line of code and describe what's happening. The words were there. The words . . ."

The judge cut her off. "I get to interrupt," she told the lawyer. "That's my privilege." Clearly not buying Epic's BS. "This was not an insignificant breach, hence, the reason we are here," she continued. "Call a spade a spade. I don't see how you can say they were being forthright in what they were attempting to do."

Forrest cited legal precedent that essentially argues that two wrongs can equal a right. In other words, Apple's contract was in the wrong, so Epic could go rogue. "Public policy dictates that when you are the subject . . . to an anticompetitive contract, you need not comply with that contract," Forrest said. "What we did here was cease complying with an anticompetitive contract. That was the sin that we committed, was ceasing to comply with an anticompetitive provision. . . . Did we know what we were doing? We did, Your Honor. We understood that to have this fight with Apple, it was going to be a big fight. It had to be done by a company that was committed to go the distance, as Epic is. Epic is a company which understood it had to prepare for this fight, which it did because it was the only way to try and break the choke-hold that Apple has on its payment system and its prohibition on any kind of competition."

To save its Unreal Engine business, Epic's argument boiled down to the notion that its fight with Apple was about *Fortnite*, while the developer tool's fate was covered under another contract with Apple that it hadn't violated. While Epic and Apple were both giving pre-views of arguments to come, on this day the real question was about whether the court would prevent Apple from blocking Epic's *Fortnite* and effectively blocking Unreal Engine. After the initial volley, the judge split the baby. She wasn't going to give temporary relief to Ep-ic's own *Fortnite* that had been booted from the App Store, and she wasn't going to let Apple harm non-Epic developers using its Unreal Engine—a decision that would stand while the matter was being lit-igated. The real case was still months away, though the judge made it clear she wanted to move unusually fast when asking when they could be ready to litigate. The Apple case was going to go before the Google fight.

"Literally, give me a month," she told the lawyers. "When are you ready?"

*   *   *

A key part of the legal battle between Apple and Epic would be over defining the market that was in dispute—what could be a very wonky and technical-filled fight. Epic would then need to show an illegal monopoly of that market. Judge Gonzalez Rogers would decide whose version of the market was correct. She also would decide the merits of the case—no jury. It would be a bench trial. The decision was likely to resonate, and legal experts suspected it could remake the nation's inter-pretation of antitrust laws when it came to digital markets.

The core of Tim Sweeney's antitrust argument was that Apple had an improper monopoly because it held sole control over how outside apps could be installed on iPhones and forced those developers to use its own payment system. With that, Epic's lawyers were preparing to argue that Apple was able to dictate anticompetitive commissions—the 30 percent tax that Sweeney was always talking about.

Unsurprisingly, Apple painted an entirely different picture. It saw the competitive landscape much larger than just apps on its iPhones but video games in general. With that reasoning, Apple argued that *Fortnite* could be played in all sorts of ways—from PlayStation machines to personal computers. When the game was pulled from the App Store, lawyers noted, less than 10 percent of average daily players were using Apple devices. Its empire made up just a small part of Epic's business, underscoring that Apple didn't control the entire video game market. "There is huge market competition," Richard Doren, one of Apple's lawyers, told the judge in a pretrial hearing as the date for the trial approached. "The market is broad. . . . And Apple does its best to compete."

He continued that using antitrust analysis, the question for the judge to decide was whether Epic had alternative channels to distribute its game. "Fortnite is not entitled . . . access to every single person on the globe," Doren said. "They only need to have reasonable distribution alternatives available, and they have that in spades."

Epic's outside lawyer Gary Bornstein took issue with Apple's read of its figures. Rather than 10 percent of daily active users being Apple players, he wanted the court to look at the number of *Fortnite* users who had linked their accounts to an Apple device—a much higher figure. But, Bornstein told the judge, he didn't want to quibble on the numbers, what he was trying to do was make the point that just because there was the possibility of playing the game on an Xbox or PlayStation didn't keep Apple from using its monopoly power to set prices on the distribution of apps on the iPhone. "Ultimately, I think it comes down to . . . you can't carry a PlayStation in your pocket," he said.

The judge cut him off. "You keep forgetting other devices."

"Well," he said, "but . . ."

"You're forgetting the Switch," the judge injected, referring to Nintendo's portable game player. "I went online yesterday, and at least in the Bay Area, you can't get a Switch anywhere. You can't get it at Best

Buy. You can't get it at Amazon . . . for at least a week. You can't get these devices. You can't get it at Target."

Bornstein tried to interject again, but the judge just continued. "Perhaps it's because everybody's stuck at home."

"Well," Bornstein said, "but . . ."

"But if they're stuck at home, they're probably downloading it on their TV's," she continued, in reference to Xbox and PlayStation machines. It was still months before the judge would rule on a market definition, but clearly Epic had more work to do to convince her. "It's just not a substitute for being able to reach the iOS users that there might be some other channels," he told her. "I don't want to lose sight of the fact that we're not talking about just games here. We're talking about an app market more generally . . ."

The judge interrupted again. "I know that that's how you want to frame the issue," she said. "I'm not convinced that you have."

More ominously for Epic, she continued, "I think this is going to be a fascinating trial. And like I said, there's great debate in the industry. . . . But it's complicated because if we look at . . . this industry, walled gardens have existed for . . . decades. Nintendo has had a walled garden. Sony has had a walled garden. Microsoft has had a walled garden. And so in this particular industry, what Apple's doing is not much different." Still, she said, Apple is more than video games. They have created a platform for general app markets—a new world. And, to underscore just how unprecedented the trial was going to be, she declared from her California-based bench, "They don't call this the 'Wild West' for nothing."

Not to be cowed by the judge's forewarnings, Epic's lawyer continued to try to challenge the judge on her view of the market, noting that the economics that fueled the business of video game machines were much different from those of the iPhone. Apple's 30 percent fee wasn't tied to the business of making phones. The question to look at, Bornstein said, was whether a monopolist was in a position to profitably raise prices above the competitive level. In Epic's eyes, Apple could set such a high price because there wasn't any competition. Apple was the only game in its Walled Garden.

Again, the judge cut him off. "Plaintiffs always want me to define relevant markets as narrowly as possible," she said. "That helps their case. And defendants always want me to define markets as broad as possible because it helps their case. I think we can all agree that the definition of a relevant market is a factual question. It's not a legal question. Right?"

Though she phrased it as a question and Bornstein tried to respond, the judge wasn't really asking. She was making a point—a point that was becoming very clear about how things would go once the actual trial began in May. "Ultimately, you've got the burden of proof," she said.

Epic had its work cut out for it.

15
SIXTEEN DAYS
IN COURT

2021

Wearing a cream-colored dress with fluffy sleeves, Margrethe Vestager looked almost angelic as she stood at a lectern within the European Commission's headquarters in Brussels, ready to deliver some divine justice. It was a day long coming for Spotify, the culmination of a plan put into place years earlier by Horacio Gutierrez to draw the European Union into its battle with Apple. Vestager was there on the final day of April 2021 to announce that the commission was charging Apple with violating the bloc's antitrust laws by abusing its control over the distribution of music-streaming apps, like Spotify. "This case is about the central role of app stores in the digital economy," Vestager told the gathered reporters. "An app store can become a gatekeeper, in particular if there is only one app store available in the mobile ecosystem."

As when Spotify first brought its complaint on the matter in 2019, Apple appeared ready with a swift public relations response, pushing back with a full-throated defense. Just like in its fight with Epic Games on the other side of the world, Apple was framing Spotify's complaints as fueled by greed—an ungrateful business partner that had benefited from the digital world it had created and was now unwilling to pay its fair share. "At the core of this case is Spotify's demand they should be able to advertise alternative deals on their iOS app, a practice that no store in the world allows," Apple said in a statement. "The Commission's argument on Spotify's behalf is the opposite of fair competition."

While those proceedings would take many months to resolve themselves, the timing of the charges could not have been better for the rebellion against Apple. In California, Epic Games' army of lawyers was preparing for the start of the sixteen-day trial to determine whether Tim Sweeney's Project Liberty was righteous. No matter the outcome of Judge Yvonne Gonzalez Rogers's ruling, all sides quietly expected the loser to appeal, and the case would likely go all the way to the Supreme Court. At stake was Epic's business, Apple's hold on the App Store, and the future of the App Economy. The outcome could also affect actions by the Department of Justice, which had been reviewing Apple's business and was watching closely. A resounding win by Apple and any precedent might scare off government lawyers, legal experts said. At the very least, it could have them reevaluate the theory of any antitrust case.

For the first day of the trial, Sweeney showed up at the Oakland, California, courthouse in a dark navy suit and tie. Gone were his trademark hiking pants. Nothing was normal about this trek. While the pretrial hearings had been conducted via Zoom, the judge wanted the actual trial to occur in person. In a first for Gonzalez Rogers, she was allowing the proceedings to be broadcast publicly because the number of people allowed in the courtroom was severely restricted to basically a few lawyers on each side and a rotating spot for reporters. Hundreds of people—including lawyers from Spotify and other app developers who had a lot at stake as well—would sign up every day to listen as witnesses testified and as the judge interrupted with her own questions. The day's proceedings were delayed when apparent fans of *Fortnite* flooded the video connection, which wasn't properly muted, shouting profanities. "Free Fortnite!" some yelled.

In another sign of how quirky the trial was going to be, on its first day, as Sweeney sat in the witness stand, his lawyer entered into exhibits as evidence various video game machines, a Switch, PlayStation, and Xbox. The judge was familiar with the Switch; her daughter now owned one—apparently a purchase made during the pandemic. Richard Doren, Apple's high-priced outside lawyer, joked that after the case ended, they could offer the machines up for a raffle. "I do have to say one of my favorite cases was a wine trademark case and I admitted into evidence about 30 bottles of wine," the judge told the lawyers. "And at the end of the case, I did send out an order to all the parties saying within the next 30 days you must retrieve all your evidence or the Court will dispose of it." On the thirty-first day, she added, "all of the law clerks came to chambers, and we promptly disposed of the evidence."

The moments of humor peppered what was clearly a tense few hours for Sweeney, whose voice, at times, was barely audible, muffled by his protective mask and his habit of talking softly. The judge would tell him to speak up on more than one occasion ("You are mumbling again"). Across the room sat Apple's Phil Schiller, Sweeney's nemesis. The two men would dutifully report to the courtroom each day for the weeks to come, sitting on opposing sides.

"It took me a very long time to come to a realization of all the negative impacts of Apple's policy," Sweeney told the judge. "In the very early days of iPhone, iPhone games were built by a team of just a few people, and to reach a large audience. And it would be very lucrative for developers. But over that period, again, as competition grew dramatically, as the number of apps increased, there were many products being developed by teams of hundreds or more, engineers or developers. The economic model had changed such that the average developer's profit margin in my experience was well under 30%. And so we got to a point where Apple was making more profit from selling a developer's app on the App Store than the developer was typically making themselves."

The CEO told the court that the Apple empire was important to Epic's future goals with evolving *Fortnite* into something bigger than just a game. "Our aim with 'Fortnite' is to build something like a metaverse from science fiction," Sweeney said. "Reaching the entire base of Apple's 1 billion iPhone consumers is a paramount goal for our company, as 'Fortnite' expands beyond being a game into this large world of the metaverse."

Sweeney's lawyer had him tackle head-on the skulduggery brought by Project Liberty, asking why he deliberately violated Apple's rules. "To show the world through conspiracy actions exactly what the ramifications of Apple's policies were," Sweeney said. "Because I felt it was very easy to mistake Apple's business model for just software distribution; whereas I wanted the world to see that Apple exercises total control of its availability of all software on iOS and that it uses that control and can use that control to deny users access to apps that have things . . ."

The judge interrupted. She wanted to know if he or his lawyers had reached out to the lawyers who were already suing Apple on the same topic in a case that she had effectively been overseeing for years. "I don't know if . . . counsel contacted them or not," Sweeney said. "I didn't. I wasn't involved in any contact."

"But you knew that there was a lawsuit already on behalf of all developers against Apple, didn't you?" the judge asked.

Sweeney: "Yes, Your Honor."

"And you just ignored that and went forward on your own case?"

"Yes," he told the judge.

Later, it would be clear that the judge interpreted this answer as Sweeney deciding to "rush the court" to protect Epic's own self-interests.

When Apple's lawyer got a chance to question Sweeney, Doren worked to counter the image Sweeney was trying to project—an idealist standing up against the giant for the greater good. Instead, Doren attempted to paint Sweeney as an executive trying to goose his own bottom line by breaking a business contract that had long benefited him. This was exactly what the outside consultants had warned Epic about a year earlier. At the end of the day, this fight could be seen as one between billionaires over royalties.

Doren quickly zeroed in on the Project Liberty hotfix as a PR stunt, noting that Epic had cut its prices for in-game purchases to customers playing on PlayStation, Switch, and Xbox while continuing to pay 30 percent commissions to those platforms. In response, Sweeney said he saw Apple and Google's 30 percent fees as fundamentally wrong while similar fees charged by video game machines were part of the fabric of the industry. "We subscribe to the idea of subsidized hardware and felt that we were a beneficiary of that," Sweeney responded.

"But there were two reasons why you lowered the prices, right?" Doren countered. "One was for a public relations campaign to back up your Project Liberty plot against Apple, correct?"

Sweeney answered that he wanted to demonstrate the savings.

"And the other was to try and create some excitement around buying . . . making more in-app purchases within 'Fortnite' because you were concerned that interest in the game was flagging over the long run, correct?"

"The first part is correct," Sweeney said, "the second part is not."

Apple also focused on the fact that more than a decade earlier, Epic willingly signed a contract with the iPhone maker agreeing to its terms about the App Store and 30 percent cut. "In August 2020, you, as a shot caller at Epic, chose to intentionally breach your contract with Apple, correct?" the lawyer asked.

Sweeney: "Yes."

*   *   *

Looking at Epic Games' witness list, Apple began to see too many coincidences. Lori Wright, a senior Microsoft executive, was scheduled to testify. It was just the latest sign to Apple executives that Microsoft was playing more than just a passive role in its ongoing fights.

Apple executives had grown increasingly convinced that Microsoft was all too willing to help Epic. Records uncovered through discovery seemed to show a close relationship between Sweeney and the Xbox team. Microsoft had come out publicly in support of Epic's fight, and then Epic was proposing to put a senior Microsoft executive on the witness stand to help paint Apple as recalcitrant.

The history of tensions between Apple and Microsoft was the stuff of legend in modern American business. Apple's co-founder Steve

Jobs and Microsoft's co-founder Bill Gates had been rivals, duking it out during the early days of the personal computer. At the core was a difference in how the two saw the digital world. Apple was a closed kingdom that would grow into the iPhone empire. Windows was open and much more like the world Sweeney was advocating. Ultimately, Gates won those early rounds as Microsoft became a dominate player in the PC game, rescuing Apple with a much-needed investment as it neared bankruptcy.

But those hostilities were a generation old and between leaders no longer at the helms. In recent years, the two companies had maintained something of a cold war. Microsoft's business had evolved in ways that it would have been hard to imagine, underscoring the fast-moving nature of the tech industry. Chief Executive Officer Satya Nadella had rebuilt Microsoft's businesses around cloud computing, strengthening its position in business-to-business sales, while Apple had become a dominant consumer product company. But fault lines were emerging: Apple had seemed to play Microsoft for leverage in its negotiations with Google. Microsoft's video game business was now clearly in competition with Apple's hold on gamers. And the two had their own developing visions for virtual reality and artificial intelligence. Like Epic, Microsoft could benefit from a world where Apple's rules didn't control entry into technology's next-generation products—whatever they might be. In that context, Apple seemed to bristle at Microsoft's fingerprints, which it saw all over the case.

Several days into the trial, Epic called Wright, the senior Microsoft executive, as one of its star witnesses to detail the challenges she faced bringing a bundled video game streaming service to the App Store. She testified that Microsoft was treated differently than how Apple had treated Netflix, which streamed videos. "A reasonable observer might wonder whether Epic is serving as a stalking horse for Microsoft," Apple said in a court filing. "Yet Microsoft shielded itself from meaningful discovery in this litigation by not appearing as a party or sending a corporate representative to testify."

In reality, the filing would have little sway in the case beyond highlighting that Apple's fight against Epic was much more than it appeared. Apple was making the case that it faced sweeping competition. And here, in Apple's opinion, was another example of how it was under attack.

*   *   *

Eleven days into the trial, Apple called Phil Schiller to the witness stand to defend the App Store that he had devoted so much of his life to building. Nine months earlier, Apple had announced he was taking a step back from running marketing, a move intended to give him a little more time with his family. He had envisioned a vacation. In reality, Schiller would come to realize that he wasn't wired to take a vacation; he didn't know how stop working all of the time. Instead, Schiller had kept with the eighty-hour work weeks. And he applied that workaholic approach to the trial. Almost like a student, he sat there in the courtroom studying the judge and learning how she thought about issues, hearing how Epic and its witnesses interpreted Apple's actions from the outside of Apple Park. Their view of Apple was not the Apple he knew. He held on to the strong belief that for most developers—and users—they loved the way Apple did things. In many ways, Schiller was there to provide a history lesson on why Apple operated the way it did, why it was different, and, perhaps, why it was successful. It was a less than subtle message to the judge. Tim Sweeney was asking the court to change how Apple did business. Judges rarely want to dictate how a business operates. "Apple always considers ourselves not a hardware company, not a software company, but a product company," he said from the witness stand.

Schiller's job that day was also to get across a major part of Apple's defense: that it controlled access to the iPhone through the App Store for security purposes. The company was spending a considerable part of its defense on detailing the steps it took to block bad actors from putting malicious software into the App Store.

To refute those claims, Epic wanted to paint the security concerns as a pretext for maintaining its monopoly control. To illustrate the point, Epic's lawyer had Schiller wade through a collection of rather blue apps—those dealing with fetishes and 3D sex positions.

Epic also wanted to show that Apple was using its dominance on the market to extract unfair profits that come from being a monopoly. To do that, Apple's accounting was put on trial. When asked if the App Store had been profitable since 2009, Schiller said he didn't know. "How is it that as the executive responsible for this major business in the country, you don't even know whether or not it's profitable?" Epic's lawyer asked. "How can that possibly be?"

Schiller: "Because it's not how I look at the business and not what I measure the team driving and what we do for our customers and developers."

"Doesn't anybody ever wonder, 'Hey, is the App Store profitable?'" the lawyer asked. "That never comes up?"

Schiller: "No. It doesn't."

"Doesn't it suggest to you that only a monopolist can ask if a major line of business is not profitable?" the lawyer responded.

"As we explained, Apple is one P&L," Schiller replied, talking about profit and loss. "It all contributes to the profit of the company, and that's how we think about it."

*   *   *

On the second-to-last day of the trial, it was Tim Cook's turn, the first time the Apple CEO had taken the witness stand in the many cases brought against the giant company over the years. He'd faced down Congress years earlier over tax issues, then again in 2020 via video-conference to defend Apple against antitrust concerns. This time he was in person, wearing a dark suit and plastic face shield. After he was sworn in, one of Apple's lawyers brought him a glass of water. Things started out slow as Apple's lawyer Veronica Moyé asked, "Mr. Cook, how would you describe Apple's mission?"

"It's to make the best products in the world that really enrich people's lives," he said with his southern drawl. It would carry on like this as Cook made the case that Apple invests billions to make products that people enjoy and fights to protect their privacy and security. He would spend about four hours on the stand, and the gist of his message was that the App Store was run the way it was run for user security.

Epic had turned up documents that showed Apple did discuss profitability of the App Store, including a meeting between Cook and CFO Luca Maestri on the topic. Cook, implausibly, tried to suggest such analysis was unusual and that Apple didn't use that kind of information for making business decisions. Yet emails mentioned "method 2" of allocating operating expenses. "So your corporate planning and financial analysis group does have at least two methods for allocation of operating expenses," Gary Bornstein, Epic's lawyer, asked.

"I think so," Cook responded. He professed not remembering the presentation. Cook would concede that the App Store was profitable. But how did he know that, if, as he said, Apple didn't keep track of such things? "I have a feel, if you will," he said.

As Cook's testimony wound down, the judge had some of the sharpest questions yet for Cook, narrowing in on how the majority of revenue from the App Store came from in-app video game purchases. "What is the problem with allowing users to have choice, especially in the gaming context, to find . . . a cheaper option?" she asked.

"I think they have a choice today," Cook responded. "They have a choice between many different Android models, the smartphone, or an iPhone. And that iPhone has a certain set of principles behind it from safety to security to privacy and . . ."

The judge cut him off, saying that an iPhone user couldn't be notified within the iOS environment that a cheaper in-game currency for *Fortnite* existed.

"If we allowed people to link out like that, we would in essence give up the . . . the return on our" intellectual property, Cook said. And as the judge tried to poke at his response, he continued to defend it. "We need a return on our IP," he said.

She pointed to banking apps, which other than paying $99 for the developer fee, didn't get charged anything for being part of the App Store. "You're charging the games to subsidize Wells Fargo," she told Cook.

He disagreed. Gamers were making purchases on Apple's platform. "People are doing lots of things on your platform," she retorted. "So, so . . ."

"But this is a digital transaction with observable change in . . . currency," he continued.

"It's just a choice of a model," she injected. Apple had made a choice of a business model.

"We've made a choice, yeah," Cook said. "There are clearly other ways to monetize and we chose this one because we think this one overall is the best way."

"Well," the judge shot back, "it's quite lucrative. But it seems to be lucrative and . . . focused on . . . purchases that are being made, frankly, on an impulse basis—that's a totally different question about whether that's a good thing or not, not really ripe for antitrust law. But it does appear to be disproportionate." She continued that she understood Apple's argument that it was bringing the customer to the games but then narrowed in on Sweeney's long-lingering gripe that after the first interaction it wasn't Apple's work but the game developer's work that kept them engaged. "Apple is just profiting off that," she said.

Cook disagreed. "I view it differently than you do," he said. "I view that we're creating an entire . . . amount of commerce on the store, and we're doing that by focusing on getting the largest audience there." Apple does that, he said, by allowing free apps to create interest.

"You would agree with the basic proposition that competition is good," the judge asked.

"I think competition is great," Cook replied. "We have fierce competition in our business."

"You don't have competition in those in-app purchases, though," she said.

"I mean somebody could go if they're . . . a gamer, they could go buy on the Sony PlayStation or the Microsoft Xbox or the Nintendo Switch," Cook said.

"Well, only if they . . . only if they know," the judge said.

"Well," Cook said, "but that's up to the developer to communicate."

The appearance was major news. An artist rendering of Cook before the judge appeared the next day on the front page of *The Wall Street Journal*. A *Washington Post* article declared that Cook had "faced some of the most forceful scrutiny of a major tech executive in decades." But would it matter?

*   *   *

With the trial complete, Judge Gonzalez Rogers had to decide who was right: Apple or Epic? It would take her several months to issue her findings. In the end, no one would be happy.

On most of Epic's case, the judge disagreed. Perhaps unsurprisingly given her comments, she didn't agree with Epic's market definition. And she didn't see Apple as violating the Sherman Antitrust Act and didn't agree in calling for knocking down its Walled Garden. But on the issue of Apple blocking developers from alerting users of cheaper options outside the Apple empire, that struck her as wrong—a violation of a California law against unfair competition.

Apple's provision barring apps from directing users to other ways of paying was one of the final topics the judge covered with Tim Cook when he appeared on the witness stand during the trial. The anti-steering provision was the issue that had tripped up Spotify years earlier in its fight with Apple and was at the core of its complaint to the European Commission. Simply put, Apple, in her opinion, was denying users choice. She disagreed with Cook that Apple was doing what other physical stores did by not letting rivals advertise within their walls. "These restrictions are also distinctly different from the brick-and-mortar situations," she wrote. "Apple created an innovative platform but it did not disclose its rules to the average consumer. Apple has used this lack of knowledge to exploit its position."

That needed to change, she wrote—ordering the biggest change in Apple's App Store since its early days. But she stopped short of calling Apple an illegal monopoly. Cook and his team had sidestepped a massive blow that would have meant huge changes to its business model. Rather, the judge ruled that Epic had broken its contract with its Project Liberty because Apple didn't have an illegal monopoly that justified its exploits. Epic, she wrote, ignored the realities that customers should be able to choose between ecosystems and that features, such as phone battery life, durability, ease of use, and performance, factor into the market. She also didn't agree with Epic's theory that Apple had improperly made its products so that customers were locked in and couldn't easily switch. "Apple sought to compete by distinguishing their product, and in the process, making its platforms 'stickier,'" the judge wrote. "That, however, is not necessarily nefarious."

She decided that Epic never showed that Apple's hold on the iPhone world was an actual market. "Creating a seamless system to manage all its e-commerce was not an insignificant feat," the judge wrote. "Under all models, Apple would be entitled to a commission or licensing fee, even if IAP was optional." Instead, she defined the market as all digital gaming transactions. "While the Court can conclude that Apple exercises market power in the mobile gaming market, the Court cannot conclude that Apple's market power reaches the status of monopoly power in the mobile gaming market," she decided. "That said, the evidence does suggest that Apple is near the precipice of substantial market power, or monopoly power, with its considerable market share. Apple is only saved by the fact that its share is not higher, that competitors from related submarkets are making inroads."

The split decision came in part because the law allowed for what Apple was doing with its App Store, the judge ruled, fueling belief among Apple critics that the law needed to be changed—that twentieth-century regulations weren't keeping up with the realities of the twenty-first-century market. Both sides would appeal. For now, the judge had left the Walled Garden mostly intact. She had weakened it with her decree that developers should be allowed to communicate other payment options outside the Apple payment system. Epic's effort would serve as a lesson for others who were thinking about their own line of attack against Apple's power—including the lawyers at the Justice Department. The biggest legacy of the case: Apple's dirty laundry had also been aired. The war wasn't over. It was just moving to other venues—now with new data points thanks to Epic—and using different tactics.

PART IV

TECHLASH

16 THE CULTURE WAR

2021

No sooner had hundreds of people stormed the U.S. Capitol on January 6 than blame started to be ascribed. What occurred that day—who was right and who was wrong—will be left for others to explore. The ramifications of that day and how big tech companies, including Apple, were perceived by many people in the United States are very much part of *this* story.

Following a speech by President Donald Trump in which he aired his grievances over losing the November election, which he claimed was stolen, supporters stormed the Capitol, leaving several people dead. Trump's role would long be debated as would social media's part. In the aftermath, Facebook temporarily suspended Trump's account, then indefinitely extended that move. Twitter followed, banning his personal account and taking down posts attempted through the president's official handle. A nascent social media network called Parler, which had positioned itself as a home for conservative users who felt unfairly censored by Facebook and Twitter, came under pressure from Apple and Google. The high-profile episode would soon frame the power Apple holds over its iPhone empire in a new way, no longer just billionaires fighting over percentages but how central the digital world was to debates raging in the real world.

Founded in 2018 by John Matze, Parler (pronounced like "parlor") came to be at just the moment when conservative users were looking for an alternative to Facebook and Twitter, which had begun policing contentious content following the 2016 presidential campaigns. After Trump's election, both platforms were accused of being used for Russia-backed meddling.

Parler was essentially a poor man's Twitter. In 2020, Matze, twenty-seven, was clearly still trying to figure things out for the nascent company. During a *Forbes* interview that summer, for example, he was asked how he would handle it if a user used the N-word in a post. He seemed to be processing his company's response in real time during that interview. First, he told the reporter the company would leave it alone. Then he changed his mind a bit later. "If somebody came on there and said the N-word to somebody, and they got very upset as a result of that, then it would get taken down."[1] In reality, however, his bare-bones operations didn't have the kind of staff or technology to take such steps at scale. He was relying on volunteers to try to help combat an influx of liberal users flooding the site to harass conservatives. Parler was built on a rather idealistic belief, the kind of libertarianism that inspired many in Silicon Valley, that free speech should allow all kinds of speech, including the ugly stuff that upsets people. "The best thing is for everyone to engage with a bad idea and shut it down through public discourse," he told *Forbes*.

That stood at odds with the conventional wisdom at the time that powerful social media networks had a civic duty to stop the spread of information deemed harmful. S. Matthew Liao, a philosopher, asked the provocative question in *The New York Times* in an opinion piece titled "Do You Have a Moral Duty to Leave Facebook?" In his essay, which name-checked Facebook's Cambridge Analytica scandal and other instances of Facebook being used to spread "white supremacist propaganda and anti-Semitic messages," he noted the perceived problem that many had: "A significant amount of fake news can be found on Facebook, and for many users, Facebook has become a large echo chamber, where people merely seek out information that reinforces their views."[2]

Social media platforms had broad protections for content posted by users under a foundational U.S. internet law known as Section 230. These companies weren't liable for harmful content posted on their sites, a shield intended to protect free speech. But they're also allowed to remove what they deem objectionable if they act in good faith. Generally, Facebook and Twitter had taken a hands-off approach. In fact, Twitter's role in fostering controversial speech had helped build its bona fides as the scourge of certain governmental regimes around the world. A decade earlier, Twitter had been a key tool in a wave of pro-democracy demonstrations in what became known as the Arab Spring.

Ahead of the 2020 presidential election, however, Facebook and Twitter both seemed to be showing signs of wobbling on their resolve. In the spring and summer, they changed their stances on how to handle posts from politicians. Both began amending President Donald Trump's posts with notes about his claims aimed at combating what they saw as misleading statements. Posts on Facebook by Trump that violated its policies around hate speech, for example, were allowed to stand with a label that said they were deemed newsworthy—an apparent compromise trying to balance CEO Mark Zuckerberg's stated belief that political leaders should not be policed because knowing their views is in the public interest with the increasing outcry of advertisers threatening to pull spending.[3]

All of this was a boon for Parler. In response that summer, several Republican lawmakers and other high-profile conservatives began putting Parler on the map as the alternative. The list of top users looked like a who's who of the MAGA movement: Candace Owens, Laura Loomer, Charlie Kirk, Breitbart News. In July, a *Politico* headline declared "PARLER FEELS LIKE A TRUMP RALLY." Parler's rally was set to grow bigger. As the presidential campaign between Trump and former vice president Joe Biden became more heated, Facebook and Twitter took more steps to limit Trump. In an episode that conservatives saw as liberal tech companies putting their fingers on the scale, Twitter and Facebook limited users from sharing an October story by the *New York Post* that was potentially damaging to Biden just three weeks before the election. Shortly after the article was published, Facebook limited distribution so it could fact-check the claims while Twitter blocked the article in part because, the company said, it contained people's personal phone numbers and email addresses in violation of its privacy policy.[4] The Trump campaign complained loudly, especially after it said the White House press secretary, Kayleigh McEnany, had been locked out of her Twitter account for posting the *New York Post* article. "This is real election interference," Senator Marsha Blackburn, a Republican, fumed on Twitter.

After Election Day in November, as votes were still being counted, both Facebook and Twitter attached labels to the president's early-morning posts claiming the election was being stolen. Facebook added a label saying that no winner had yet been projected. Twitter put a label on the tweet that cautioned his claim was "disputed" and "might be misleading about an election or other civic process." Twitter took the added step of restricting users to like and share the post, limiting, in theory, the ability for Trump's claims to spread.[5] As votes were counted, Twitter continued labeling Trump's tweets—and, in some cases, hiding them—as misleading as well as those in his inner circle as the president fought Biden's subsequent victory. Some watched and felt the California tech companies were singling out conservative voices, igniting a huge surge of users signing up for Parler.

The Fox Business anchor Maria Bartiromo and other high-profile conservative voices began directing followers to Parler. "This is the same group who abused power in 2016," she told her Twitter followers in early November 2020. "I will be leaving soon and going to Parler. Please open an account on @parler right away." Soon, Parler ranked as the top download on Apple's App Store, almost doubling to eight million users in a week.

As Biden's Inauguration Day drew near, the rhetoric grew heated on Parler. So much so that the company began secretly sharing some alarming posts with the FBI that suggested certain fringe groups were planning violence in Washington around a planned Trump rally on January 6, as records that would become public later showed. Posts included "NO MORE MARCHING ACTION TIME" and "Hang EVERY TRAITOR ON JAN 6TH!" Another post from January 2 shared with the FBI read, "This is not a rally and it's no longer a protest. This is a final stand where we are drawing the red line at Capitol Hill . . . don't be surprised if we take the #capital building." It would be those kinds of posts that would taint Parler in the days—years—after the January 6 riots as debate intensified over social media's role in policing speech. Unlike Facebook and Twitter, which seemed, in part, to be reacting to market conditions (advertisers pulling spending), Parler didn't face such pressure. It was still struggling to figure out a business model based on its free speech principles. Rather, it was vulnerable in ways that Spotify, Epic Games, and others in the App Economy knew all too well: the rules of the app stores. Instead of fights over payment systems, Parler was falling under edicts by Apple and Google about content within its app. Just as Apple could cut off *Fortnite* because it didn't like something Epic had inserted into its game (a payment system that violated its rules) or delay Spotify updating its software because it objected to its messaging (efforts to steer users around Apple's payment system), Apple could ban Parler for its handling of content moderation, cutting off the startup's access for acquiring new users in the iPhone empire. On January 8, Apple sent Parler a stern warning: "We require your immediate attention regarding serious App Store guideline violations that we have found with your app, Parler. We have received numerous complaints regarding objectionable content in your Parler service, accusations that the Parler app was used to plan, coordinate, and facilitate the illegal activities in Washington D.C. on January 6, 2021 that led (among other things) to the loss of life, numerous injuries, and the destruction of property. The app also appears to continue to be used to plan and facilitate yet further illegal and dangerous activities." Apple required Parler to submit an update to "moderation" efforts to avoid being removed from the store. A few hours later, Google, which had long followed Apple's lead in how it managed its own app store, sent a similar message to Parler. It was as if Apple's original warning gave cover for the rest of tech to respond. Soon, Amazon.com was warning that it would boot Parler from its servers, which would effectively make the site go dark. Matze faced uncharted territory: The gatekeepers to the App Economy were threatening to ban an app because of news events playing out on its social media network. No apps, no web pages; Parler was left scrambling to stay alive.

As examples of offending conduct, Apple pointed to posts on Twitter that included screenshots from Parler. Some of the Parler posts appeared to call for more violence. Others seemed to be cheering on the activists in Washington—a digital peanut gallery. "Twitter needs stormed. Blatantly removing the tweets of the President," one user wrote, apparently suggesting that the social media company needed to be attacked after taking down Trump's tweets. Another apparently voiced support for Trump, writing, "Even if he starts killing thousands of Americans, I trust his judgement. I know they would deserve it. God made him. He can't be wrong." Those screenshots were among several collected by a Twitter user called Sleeping Giants, advocating for "social platform accountability" with a large following. The account had tagged those posts at Apple and Google the day after the attack with its concerns: "Honest question for @AppStore and @GooglePlay. If Parler continues to allow incitement and calls for violence, doesn't that break your Terms of Service for apps?"[6]

In its message to Parler, Apple cited a statement that Matze had made to *The New York Times* about concerns raised about the platform. "I don't feel responsible for any of this and neither should the platform, considering we're a neutral town square that just adheres to the law," he told the *Times* in an apparent reference to Section 230, which shielded social media platforms. That may be true in real life, but Apple saw Parler's obligations differently within its iPhone empire. "We want to be clear that Parler is in fact responsible for all the user-generated content present on your service and for ensuring that this content meets App Store requirements for the safety and protection of our users."[7]

On Parler, Matze stood defiant. "We will not cave to pressure from anti-competitive actors! We will and always have enforced our rules against violence and illegal activity. But we WONT cave to politically motivated companies and those authoritarians who hate free speech!" He continued, "Anyone who buys an Apple phone is apparently a user. Apparently they know what is best for you by telling you which apps you may and may not use."

The next day, Parler was pulled from the App Store and Play. Soon, it went totally dark when Amazon pulled it off its servers. Before going dark, Matze made one last stand, claiming Apple, Google, and

Amazon were working in concert to quash his company: "You can expect the war on competition and free speech to continue, but don't count us out."

High-profile conservatives largely agreed, adding to the voices condemning the move. But so did the American Civil Liberties Union. Ben Wizner, an ACLU lawyer, told *The New York Times* that the actions to pull down Parler went beyond Facebook and Twitter curtailing users' posts. "I think we should recognize the importance of neutrality when we're talking about the infrastructure of the internet," he said.[8]

Representative Devin Nunes, a Republican from California, went on Bartiromo's popular *Fox News Sunday* morning show to blast the tech companies. "The fact of this is that there is no longer a free and open social media company or site for any American to get on any longer because these big companies—Apple, Amazon, Google—they have just destroyed what was likely, Parler is likely a billion-dollar company. Poof, it's gone. But it's more than just the financial aspect to that. Republicans have no way to communicate," he said. "If you're Republican or conservative, if you don't want to be regulated by left-wingers that are at Twitter and Facebook and Instagram, where you get shadow banned—nobody gets to see you. They get to decide what's violent or not violent. It's preposterous."

And if that wasn't enough, he added, "I don't know where the hell the Department of Justice is at right now or the FBI. This is clearly a violation of antitrust, civil rights, the RICO statute. There should be a racketeering investigation on all the people that coordinated this attack on not only a company, but on all of those like us, like me, like you, Maria, I have three million followers on Parler. . . . I will no longer be able to communicate with those people, and they're Americans."[9]

And like that, Apple—with all its best intentions of protecting its brand and iPhone empire—was being lumped in with the rest of Big Tech censoring speech in America. It had done something that few could do in a polarized America: unite both Democrats and Republicans around the idea that it held too much power. Republicans saw it as limiting speech; Democrats as limiting competition.

*   *   *

The next battle for Apple moved to an unexpected front: North Dakota.

Epic Games, which was about to meet Apple in federal court for its sixteen-day trial, and the Coalition for App Fairness hired a local Bismarck lobbyist to take their fight that winter to the legislature as part of a broader campaign rolling out at statehouses across the United States that left Apple scrambling to hire its own lobbyists to counter. Legislation was introduced to prevent Apple from forcing apps to use its payment system and to allow apps to be downloaded outside the App Store—key tenets of Epic's fight. Soon, Apple was trying to set up Zoom calls with state senators in a bid to convince them that State Senate Bill 2333 "threatens to destroy iPhone as you know it," as Apple's chief privacy engineer, Erik Neuenschwander, testified before a senate committee that February.[10] He explained that the App Store was designed as an integrated feature of the iPhone, not a separate component. "For example, right now, your iPhone is designed to prevent software from obtaining unauthorized access to your camera or your photos or your location," he told them. "But if you force other software onto iPhone, as Senate Bill 2333 might do, you would undermine the privacy, security, safety and performance that is built into iPhone by design." Much of the case Neuenschwander made before the senators was one that would be made in the federal court before Judge Yvonne Gonzalez Rogers in May that same year in its case against Epic Games: how Apple took great pains to run its App Store to protect users, reviewing 100,000 app submissions each week, rejecting 40 percent that failed to meet the company's exacting standards.

Except before the lawmakers, Neuenschwander delivered a more folksy pitch for the digital world. "I understand some of you have owned stores yourselves, so this will be familiar to you: you don't put just any product on your shelves; you stock your shelves only with products that meet your standards for safety and quality," he told the lawmakers. "You don't want to sell products that don't work or pose a danger to your customers. And that's how we run the App Store: to keep out apps that would steal your banking information, or break your phone, or spy on your kids."

The lobbyist had persuaded state senator Kyle Davison to introduce the bill. The level of attention surprised him as the tech industry's focus turned to Bismarck. "She said to me that this could be big. But to me, that means the local newspaper is going to come with a camera," Davison, sixty, told *The New York Times* of the lobbyist who approached him.[11] "I would not be truthful if I said I expected the reaction."

Debate followed the chamber's decorum rules, which prevented senators from using Apple's formal name. Instead, Apple was referred to as a "technology company" or, in one instance, "the same fruit Adam and Eve" were asked not to eat.

Opposition seemed overwhelming.

"North Dakota is not the place to settle a dispute between companies on what the commission rates or payment systems should be," Jerry Klein, a state senator, said.[12]

The bill failed to move past the state senate. But that marked just the beginning of a long slog for Apple, because similar bills had been introduced across the country, including in Georgia and Arizona, just as the nation's political mood about Apple appeared to be shifting.

*   *   *

In Kingman, Arizona, a dentist turned state lawmaker named Regina Cobb, a Republican representative, also got a call from a local lobbyist hired by the Coalition for App Fairness asking if she might sponsor a bill in the house.

It was essentially identical to the effort in North Dakota. Once the Republican filed it, all hell broke loose.

Cobb told the *Arizona Mirror* her phone blew up. "It was nuts," she said, "I had several lobbyists call me and text me over the weekend." Apple and Google hired the equivalent of special forces to kill the bill, including the governor's former chief of staff. In the span of a few months, Apple tripled its efforts from two lobbyists at one firm to six at three outfits. It brought in Tim Powderly, its own Washington-based government affairs executive. The lobbyists used a similar playbook with lawmakers as in North Dakota. One Apple lobbyist even forwarded to a house committee chairman a letter written for North Dakota senators by the American Legislative Exchange Council, saying the bills were so similar "the sentiments are likely to be the same." In that letter, the president of the conservative group's advocacy arm ALEC Action, Michael Bowman, seemed to allude to rising concerns about the App Store being used to censor content. "There is no need to rewrite antitrust doctrine to protect online speech, and doing so will likely harm consumers," he concluded.[13]

The anti-Apple side came out in full force, too, including David Hansson, co-founder of Basecamp, the maker of an email productivity app that had fought App Store rules. "We almost lost our new email business last year when Apple showed up with a baseball bat and demanded we hand over 30% of our revenues or they'd bust our kneecaps (i.e. throw us out of the store)," he told the chairman and committee members in an email. "So for me, this is very personal."

He laid out how his company has fewer than sixty employees and took a "huge gamble" spending millions of dollars to create an email service to rival Apple's and Google's offerings.

"When those platforms form a duopoly on distribution of mobile software, we can't just leave them to do whatever they want with that duopoly," he told them. "Apple and Google will both say 'Oh, but you don't need to use our app stores, you could also just make software for the web,' which is tantamount to the railroads telling farmers and manufacturers that they don't need to use their rails because they could always ship their products on the back of donkeys!"

Cobb thanked him for his support, confiding that she was facing "the heat" not to hear the bill. "Thank you," he told her, "for not letting Apple and their army of lobbyists bully you or the rest of the committee from hearing this issue."

"I am not easily bullied," she assured him. That day, she pulled off the unexpected, getting the legislation passed through a committee by one vote, sending it to the broader house for consideration. There, it won Republican support, getting a 31–29 vote to send it to the senate, where it would die as lobbying intensified.

The vote in the house was informative. Statehouses were often called the laboratories of democracy where new ideas are tested. What the Arizona bill showed was that rank-and-file Republicans, the grass roots of the party in an important swing state, were against Apple's unquestionable control of its iPhone empire.

*   *   *

The House's investigation into Big Tech, which included the dramatic appearance of Tim Cook and rival CEOs, via video link in 2020, had resulted in a blistering report about perceived abuses by Amazon, Facebook, Google . . . and Apple. In October of 2020, the Democrats on the House Antitrust Subcommittee, capping a sixteen-month investigation, released the 449-page findings. "To put it simply, companies that once were scrappy, underdog startups that challenged the status quo become the kinds of monopoles we last saw in the era of oil barons and railroad tycoons," the report said. When it came to Apple, the report concluded Apple exerted monopoly power in the mobile app store market, controlling access to more than 100 million iPhones and iPads in the United States. The question, in early 2021, was next.

A sweeping bill aimed at the four tech companies was being crafted in the Senate, but it wasn't clear yet that it could gain enough support to be passed into law. It was so large that the idea of it was surely going to attract naysayers. That winter an article in the Fargo *Forum* about the proposed legislation against Apple in the North Dakota legislature caught the attention of Senator Richard Blumenthal's office. Blumenthal, a Democrat from Connecticut, had long been involved in causes trying to hold big corporations to account. As state attorney general in Connecticut years earlier, Blumenthal had helped kick off the case against Apple over its part in the e-book antitrust case. As a senator, he was part of efforts to change the nation's antitrust laws.

Blumenthal's staff had taken note when months earlier, Epic Games gained attention when its popular *Fortnite* game got booted from Apple's App Store. That was tangible. Players could no longer download the game on their iPhones. And Epic had shown how it could be cheaper, too. Maybe this was an issue—narrowly defined unlike other more sweeping antitrust proposals—that could find consensus?

By April, the Senate's subcommittee on antitrust was holding a hearing on app store competition—hauling Apple and Google executives before it for questioning. And it quickly began with a display of growing bipartisan support of outrage at the behavior the senators were seeing. It had almost become expected that certain Democrats would speak out about the power of Big Tech. But Senator Mike Lee, a Republican from Utah, made it clear he wasn't happy with how Parler was taken down. "The power that Google, Apple and a few other large tech companies hold over the way Americans live their lives is, itself, simply unprecedented and legislators on both sides of the aisle are right to be concerned."

He noted that Parler was taken down "despite Parler's close collaboration with the FBI in advance of the January 6 events in order to flag potential events" and despite "the much greater prevalence of planning of the horrible events" on Facebook, Instagram, and Twitter. "Those apps, you know, were allowed to stay on while Parler, disfavored by the woke barons of Silicon Valley, was just taken down—millions of voices went silent, and one potential avenue for competition in this marketplace for social media platforms, it just disappeared into thin air," Lee said. He noted that he reached out to Apple almost immediately. "Apple's story seemed to change somewhat over the course of our discussions," Lee said, noting that he understood a resolution had been worked out to see Parler return to the App Store. "I am very dismayed that it took three and a half months and the intervention of a United States senator in order to get there," Lee continued. "It's dumbfounding to me, these are not the actions of companies that feel like they have meaningful competition. . . . It does make we wonder a number of things, how many small businesses—and let's be clear, that's what many apps start out as— . . . flounder and die because of potentially pretextual restrictions imposed on them arbitrarily by the App Store gatekeepers."

He concluded that Parler's removal might not be a violation of antitrust law, but "it might be symptomatic of the immense market power held by a handful of tech companies. . . . When we see this kind of market power being used in harmful, if legal ways, it's only fair to ask whether it's also being used in illegal ways that might hurt . . . competition. . . . If big tech is going to take a side in the culture wars or in political conversations, big tech should be prepared for greater scrutiny that will come with that unfortunate choice and with that harmful choice."

Apple and Google executives, who defended their companies actions in general, were joined by members of the rebellion against them. That showing included Spotify's Horacio Gutierrez, the architect behind the streaming music company's global campaign against Apple, and Tile's general counsel, Kirsten Daru, who was there to talk about how the company that sold tracking devices that connected wirelessly to iPhones to keep tabs on keys, wallets, and other items felt like Apple was preferencing its own late entry into the space.

The real bombshell, however, came from Match Group's top lawyer, Jared Sine. Senator Amy Klobuchar, the Democratic chair of the subcommittee, turned to him with a question about whether companies faced retaliation for speaking out, noting she had heard from companies afraid to testify.

"We have had outreach with respect to our coming forward," Sine said.

"By who," she asked.

"By Google."

"What happened," she asked.

"They called us last night after our testimony became public to ask us why our testimony was different than what we said about the situation in our earnings call earlier this year," Sine told her.

"Is it different," she asked.

"What we said in our earnings call earlier this year was that we believe we would be able to work through the issue of Google imposing their 30% on us, which we have been working very hard at over the last few years, meeting with regulators and others to try and change these practices and ensure this from not happening," he said.

The senator then asked if Google had the power to hurt Match. "I just want to know . . . could they hurt you in little ways that it would be really hard to detect?"

"They could hurt us in little ways. They could hurt us in big ways," he responded. "They could easily remove our app. . . . We're all afraid."

The senator turned to Google's representative, Wilson White, senior director of public policy and government relations. He chalked the call in question up to a business development team member simply calling to ask why Match's testimony, filed ahead of the hearing, was different than what the company had said in an earlier public call with investors.

"I respectfully don't view that as a threat and we would never threaten our partners," White told the senator. "Senator, having developers choosing to distribute through Google Play is core to our business, so the idea that we would threaten partners who actually share in the success that we have as a platform . . . is the antithesis of how we carry out business."

The process of making laws is one that is, in theory, devoid of emotion. But there are, perhaps, fewer things that anger lawmakers like the suggestion of a witness being intimidated from testifying before them. That exchange would be searing, according to those familiar with the subcommittee's work. By summer, Blumenthal would author a bill, cosponsored by Klobuchar of Minnesota and Republican Marsha Blackburn, a senator from Tennessee, that specifically targeted Apple and Google and aimed to limit their power over their app stores. It was similar to the legislation being pushed by Spotify and Epic moving through statehouses around the country. Blumenthal's bill was narrowly tailored, as its name suggested, the Open App Markets Act, and it was aimed directly at Apple and Google app stores. It would allow sideloading and prevent anti-steering provisions, like the kind Judge Yvonne Gonzalez Rogers would order stopped later that year in the case brought by Epic against Apple. Initially, the bill would get less attention than Klobuchar's more sweeping bill, introduced in the fall of 2021, aimed at all the big tech companies, the American Innovation and Choice Online Act.

By 2022, roughly a month after endorsing Klobuchar's signature bill, the full Judiciary Committee held a hearing on the smaller, more focused Open App Markets Act. For the first time in the new big tech era, Apple felt the full brunt of Congress. As supporters discussed why they liked the bill, a theme emerged. Democrats saw it protecting competition while some Republicans saw it protecting free speech.

A few months prior to the committee's February vote, Tim Cook attended a tech conference in Salt Lake City to share a stage with Lee, the Republican member of the Judiciary Committee who was unhappy about Parler being taken down for a few months. It was a rare appearance for Cook, who limits most of his public speaking to Apple's tightly controlled events. But the stakes for Apple were becoming increasingly high. Onstage, Cook warned Apple shouldn't be lumped in with social media companies—an apparent push back to his company being grouped into the larger big tech bill that included provisions aimed at Facebook. "The industry isn't monolithic," he said. "They're very different segments and very different markets . . . we're not in the social-media business."

At the time, Lee praised him onstage. But when it came to vote in committee, he cast his vote for advancing the app store legislation, citing Parler's removal from the app stores as part of his motivation. He noted the companies were "wielding an extraordinarily large amount of market power."[14]

During the February 2022 hearing, Blumenthal, the Democrat from Connecticut, was especially blistering about the power the app stores held. He picked up on the app developers' argument about likening Apple and Google to railroad monopolies who own the means for other companies to get their products to market. "Google and Apple own the rails of the app economy as much as the railroad companies did at the start of the last century," he said. "If you're a consumer, what this measure means to you is cheaper prices, more innovation, better products and more consumer safeguards by opening the walled garden."

At the end of the day, the legislation moved out of committee with overwhelming support, 20–2. The reaction was a stunning rebuke of Apple, even if the bill still had a long and uncertain road ahead.

*   *   *

As Elon Musk walked into Twitter's headquarters in late October 2022, he probably didn't realize that he wasn't the star, but rather now a supporting cast member of a ragtag group of rebels fighting Apple for control of the internet—all of whom were engaged in an increasingly political fight. To many, the purchase of Twitter had seemed like a joke. But his latest business venture put him squarely into the culture wars embroiling Washington and Silicon Valley—a place that just a few years earlier seemed to be the bastion of Democratic Party power, but was showing growing signs of a libertarian backlash. To Musk, Twitter was his favorite place to joke, bicker, and kill time between launching rockets and designing cars, and he increasingly worried that it had jumped the shark, becoming too beholden to the establishment Left of the Valley. In his mind, what better civic duty than to return Twitter to a town square of free expression, a forum for all speech, including the most offensive things one could say?

Musk had become aghast at how Twitter and Facebook had tied themselves in knots to pull down content that some found objectionable. Those companies had been under great pressure to rein in hate speech and misinformation around COVID-19, and more pressingly to combat U.S. election interference following the 2016 presidential elections that involved Russian hackers and social media propaganda.

The problem for Musk and a lot of people who saw the world as he did was that whatever their intentions—whether on speech, COVID, or political interference narratives—the Big Tech companies were not always right. And in fact, they could be wrong. To Musk and the like-minded, the efforts of Big Tech to silence alternative opinions in the face of establishment failure had a whiff of totalitarianism. Moreover, it was reflective of a trend that a faction of Silicon Valley "doers" despised: A new generation of left-leaning bureaucratic hangers-on had become the voice of the Valley. And that voice was a parrot who didn't actually know anything or do anything. They were the vapid and spoiled children of the world the Valley's founders had created.

When it came to Musk's beloved Twitter, what had resulted was a whack-a-mole world where content moderation played editor on the platform, which had been founded on the ideals of being a place for quick, direct, and unfiltered messaging. And, if Musk were being honest with himself, the most glaring examples to Musk seemed to penalize Republicans at the behest of Democrats. If Twitter could de-platform a sitting president, whose voice was next?

Some might argue that this is precisely what Twitter is meant to do—and has every right to do. But Musk loved Twitter because he felt it was precisely *not* meant to do that. And its direct-to-audience power had helped him immeasurably in crafting narratives surrounding his work and gaining support for his companies. One doesn't have to be particularly concerned with Musk's coherence or lack thereof to appreciate that his vision for Twitter ultimately heightened the pressure on Apple.

Like so many tech companies, Twitter was largely dependent on Apple for users to access its app through a simple push on the screen of the little blue square with the white cutout of the company's bird logo. More than 60 percent of U.S. Twitter users came to the social media app through their iPhones and iPads. Those users also tended to be richer and more desirable to merchants than those who came from the other option, Google's Android.

Spotify, Epic Games, and others had long complained about Apple's cut. Now, that fall of 2022, Musk seemed to be picking up the mantle, joining that loose group of tech companies willing to stand up to Apple. Even before buying Twitter, he had signaled that support, tweeting out in July 2021, "Apple app store fees are a de facto global tax on the Internet." His vision for Twitter's business model would ultimately place him in Apple's crosshairs—in part because he spoke freely about the potential of it becoming a Super App. "In China," Musk told supporters ahead of purchasing Twitter, "you do everything in WeChat. Outside of China, there's nothing like it, people live on one app. My idea would be like how about if we just copy WeChat?"

But more important, Musk now seemed to see the control over content as something more sinister. Outsiders might ask, Was the censorship fight really about a freewheeling Twitter, or was it actually about Musk's Super App aspirations?

As owner of Twitter, Musk's complaints about the App Economy took on new weight and attracted more attention. "App store fees are obviously too high due to the iOS/Android duopoly," Musk posted that November. "It is a hidden 30% tax on the Internet." His comments were coming at a chaotic time for Twitter—a period that saw Musk slash a large number of employees and change policies around content moderation. Advertisers were under pressure to boycott Twitter amid concerns that Musk was allowing hate speech to flourish.

A few day after Musk's duopoly tweet, Phil Schiller, the so-called App Store chairman, deleted his Twitter account and its more than 200,000 followers that he had built up since 2008. But it looked like more than Schiller voting with his thumbs. According to Musk, Apple had quietly delivered a message that Twitter could soon find itself booted from the App Store—presumably over content moderation issues. Never one to bow from a fight, Musk quickly went on the offense, declaring "war" on Apple. He went public with Apple's threats, saying the company had threatened to take Twitter off the App Store, though he didn't say why. Such a move by Apple would be silencing more than 350 million Twitter users. "Apple has mostly stopped advertising on Twitter. Do they hate free speech in America?" Musk began in his tweetstorm. "What's going on here @tim_cook?"

He continued: "Apple has also threatened to withhold Twitter from its App Store, but won't tell us why." In another tweet, he asked, "Who else has Apple censored?"

With that statement, made through his Twitter account to more than 120 million followers, Musk had joined the league of rebellious tech companies battling Apple and laying siege to the company's alleged monopolistic power. In taking his beef public, Musk had been able to do something that Tim Sweeney and Daniel Ek failed to do: frame Apple's power in terms that resonated beyond the business pages and courtrooms. This wasn't a fight over percentages. This was a fight over freedom of speech.

And unlike Spotify and Epic, Musk's call for war was almost immediately welcomed by powerful U.S. politicians who saw Apple not just as a gatekeeper for commerce but as a Big Brother–ish silencer of speech. Florida's governor, Ron DeSantis, a Republican eyeing a potential run for president, blasted Apple in a press conference, saying banning Twitter from the App Store "would be a huge, huge mistake, and it would be a really raw exercise of monopoly power." Senator Marsha Blackburn joined in with a tweet of her own: "Apple and Google don't like my Open App Markets Act because it will dissolve their monopolistic power. That should tell you everything you need to know about their priorities."

*   *   *

Elon Musk's outburst in late November was needed fuel to Senator Marsha Blackburn's effort to move the Open App Markets Act forward. Behind the scenes, she and her Democratic co-sponsors had been trying to build new momentum to overcome the pressure campaign Apple was placing on Congress. The app bill along with the more sweeping legislation aimed at Big Tech broadly was facing intense lobbying. *The Wall Street Journal* estimated more than $100 million had been spent by the tech industry on advertising to fend off the proposed regulations. Supporters of the bills had hoped for a summertime vote on the bigger bill, but that fizzled, as some Democrats worried about greater tech spending ahead of the November elections.

After years of mostly avoiding Washington, Apple had grown especially adept at the dark arts of lobbying. It had boosted its spending on lobbyists, including hiring some real heavy hitters. It also was largely funding a trade group supposedly set up for small developers called the App Association, according to a Bloomberg News report that indicated more than half of its contributions came from Apple.

As Musk made noise, Apple quickly invited him to Apple Park for a meeting with Tim Cook. While Apple had long avoided giving in to developers making public demands, Musk was a different beast. The richest man in the world, quick to anger and unafraid of long, public fights, he was essentially Apple's worst nightmare. Even if they were right, they faced being framed as villains by a man who had just arguably overpaid for Twitter as a vanity project because, Musk said, he wanted to defend Twitter as a free speech platform at the very moment that some lawmakers were pushing for legislation to rein it in, partly out of concern it held too much power over speech. His fellow rebels watched cautiously. He could help their cause greatly, but he also was a lone wolf.

As Cook and Musk met, the Apple CEO expressed concerns about advertising on the platform given worries about increased hate speech.

But he told Musk Apple wouldn't pull its ad spending after all—at least not then. In truth, Musk was probably more concerned about the inflow of ad dollars for the business that was rapidly losing sponsors. If Apple abandoned ship, other brands were sure to follow. He emerged from the meeting victorious. "Good conversation," Musk tweeted. "Among other things, we resolved the misunderstanding about Twitter potentially being removed from the App Store. Tim was clear that Apple never considered doing so."

Meanwhile, Blackburn was busy making last-minute tweaks to the legislation aimed at winning more support from those worried about free speech, removing a loophole that some feared would let Apple and Google censor apps in the name of "digital safety." The wide-open term worried some who argued it could be used as pretext to take down the next Parler.

Despite opposition, supporters believed they had enough votes to pass the legislation. The *HuffPost* wrote a blistering headline that lay blame on the Senate majority leader, Chuck Schumer, a Democrat from New York, that read, SEN. CHUCK SCHUMER CAN PASS LANDMARK ANTITRUST BILLS RIGHT NOW. WHY ISN'T HE?

As December wound down, Senator Richard Blumenthal, who had co-sponsored the app bill, admitted publicly what had become obvious on the Hill. Apple and other tech giants had won. "I am deeply disappointed & frustrated that Congress is poised to end the year without passing . . . the Open App Markets Act," he tweeted. "Big Tech's behemoth sway stymied our efforts," he continued in a string of tweets. "With armies of lobbyists & vast troves of smear money, they flooded the field with misinformation & played both sides."

17

INFINITE LOOP(HOLES)

2021

The playing board was becoming very confusing. Apple was winning some of its antitrust fights, but more and more jurisdictions were emerging as foes. The iPhone empire was being chipped away at—creating the risk that its grip could slip further. Judge Yvonne Gonzalez Rogers had ordered Apple to allow developers to provide links to websites outside its apps—where, in theory, those developers could charge less money for their services outside Apple's cut. The iPhone maker would manage to avoid Congress passing landmark legislation that would have dramatically remade its App Store business, but in late 2021 it still faced a threat in Brussels, where lawmakers were considering something similar. There were new threats to the business in other markets where regulators—encouraged by Spotify, Epic, Match, and others—were unhappy with its business: the Netherlands, Germany, India, Japan, South Korea, and the U.K. "Millions of us use apps every day to check the weather, play a game or order a takeaway," Andrea Coscelli, chief executive of the U.K. Competition and Markets Authority, said when launching its probe. "So, complaints that Apple is using its market position to set terms which are unfair or may restrict competition and choice—potentially causing customers to lose out when buying and using apps—warrant careful scrutiny."

Inside Apple, Phil Schiller was overseeing work on how to respond to the growing threat. Apple wasn't going to simply open the doors to its Walled Garden; it was going to use every lever possible to encourage users to remain within its embrace. Among those tapped was Rafael Onak, a manager who worked on UX. It was a role he had held for several years, working to craft the experience users have with Apple products. At Apple, in particular, that was an important mission, because the company strived to create products that were intuitive and easy to use. His role dealt heavily with communicating messages to users so they would feel comfortable. His new assignment for Schiller, however, was not about surprising and delighting users. Rather, it seemed, he was being instructed to scare and frighten. His job was to come up with ways to communicate with users thinking about clicking on an app link that would take them outside the iPhone empire. It might be the developer's app, but App Store rules could force how those partners communicated the change. Apple's secret work was code-named Project Michigan.

A presentation created by the team looked at various "warning options"—potential options for what users would be first shown as part of the link out of the app if they clicked on it. One option was a full-screen takeover with a message that the user was going outside the app into a browser. Another option was less extreme with just a small pop-up message. The third was basically just the link. Ultimately, they decided to go with the full-screen takeover. That fall Onak and his colleagues debated what the message should say. "Continue to the developer's website" didn't capture what was happening in Apple's eyes, since they couldn't be sure that's where the link was actually taking users. A better message might be "Open external website." As they continued to workshop options, they came up with "By continuing on the web, you will leave the app and be taken to an external website." Onak liked that one. "External website sounds scary so execs will love it," he wrote to his colleagues. Another colleague suggested adding the website's URL to make it look scarier.

As they continued to discuss what should be communicated to the user, the group wanted to indicate why it was beneficial to stay inside the iPhone empire and using its own in-app payment system rather than some developer's website. One idea centered on suggesting a user could pay quickly through Apple's system as opposed to how cumbersome it could be otherwise. "I think personally that is why I wouldn't bother," Joe Phillips, the design manager for the App Store, told the group. "More steps, have to find my card, type it all out, then giving another company my details." Phillips added, "to make your version even worse"; he suggested including the developer's name rather than the app name. That was a winner of an idea. "Ooh, keep going," Josh Elman, the product manager, told the group.

*   *   *

For brief moments in the late summer of 2021, it looked as if the seemingly impossible task of taking on the combined power of Apple and Google was actually within reach for Spotify, Epic, and others. First, Dutch regulators that August found that Apple was abusing its power over the App Economy when it came to online dating sites, ordering the iPhone empire to allow third-party payment options. The ruling came after a 2018 complaint by Match Group, the online dating company known for apps such as Tinder. Fees to Apple and Google were the company's biggest expense, around $500 million annually. The company had joined Spotify and Epic Games in its global effort to pressure regulators for change, including the efforts to help lobby U.S. statehouses for changes to state laws earlier that year. The August decision by the Netherlands Authority for Consumers and Markets ordered Apple to stop. "Apple must adjust its conditions in such a way that, with regard to their dating apps that they offer in the Dutch App Store, dating-app providers are able to choose themselves what market participant they want to process the payments for digital content and services sold within the app," the authority said in a statement. Plus, Apple must allow dating apps to steer users outside the app for making payments—a key objection Judge Yvonne Gonzalez Rogers had in the Epic-Apple case in California. The Dutch authorities came to their conclusion through a narrowly tailored view of the market for dating apps, which make their money through in-app payments to upgrade for premium services, such as "superlikes." Essentially, their argument was this: Unlike, say, video games, there wasn't an alternative place to go. "For these providers, offering an app is critical since consumers use dating services primarily on their smart mobile devices, and the consumer prefers using apps," the findings said. "Apps are appealing because, in that way, several functionalities specific to smart mobile devices can be used that are available in apps, but that are not available (or available to a lesser degree) on mobile websites." Key features, the authority argued, such as push notifications, GPS, and quick responses. "For dating apps, this is very important," the report said. Apple was given two months to comply, but the effort was put on hold as the company appealed the ruling.

Also in August, but on the other side of the world in South Korea, lawmakers were addressing concerns about the power of Apple's and Google's in-app payment systems. It was an issue that had garnered national attention after Google surprised developers in late 2020 with the announcement that it was enforcing its in-app payment requirements beyond just video games, such as *Fortnite*, to other digital services, such as subscriptions offered through Spotify and Match. While such rules had been the norm for Apple users globally from early on, the change was especially noticeable in Korea, where the market was dominated by Google's Android operating system, thanks to the hometown tech giant Samsung. The phone maker, with competing app store ambitions of its own, had played a key role in helping Epic Games' chief executive, Tim Sweeney, attempt to weaken Google's digital realm, a battle that ultimately escalated with Sweeney's secret Project Liberty and legal fight in California.

Suddenly Sweeney's, Match's, and Spotify's fights were very real in Seoul. Soon, national lawmakers were working on legislation—dubbed in the local media as the "anti-Google law"—to rein in the power of Apple and Google. When it passed in late 2021, it was the first global law specifically to limit how app stores operate. "Korea is first in open platforms!" Sweeney cheered on Twitter. "This marks a major milestone in the 45-year history of personal computing. It began in Cupertino, but the forefront today is in Seoul."

The amendment to the country's Telecommunications Business Act prevented large app stores from requiring developers to use their in-app payment system and aimed to protect developers from retaliation by banning companies, such as Apple, from unreasonably delaying the approval of apps or deleting them from their stores. Failure to comply could mean fines of up to 3 percent of a company's South Korean revenue. The legislation joined a growing effort by regulators around the world, sparked, in large part, by Spotify in Europe to address concerns about the dominance of Apple's App Economy. The Korean law was increasingly being seen as a bellwether. "As bills with similar implications are being proposed in the U.S. and Europe, South Korea's bill will become a cornerstone for legislating app market platform regulations world-wide," Han Sang-hyuk, chairman of the Korea Communications Commission, said.[1]

That fall, Japanese regulators agreed to drop their investigation into Apple's App Store as part of an agreement by the company to allow media apps, such as Spotify and Netflix, to provide links to sign-up pages on those companies' websites. What that change would be like in practice wasn't clear, but the actions of the Dutch, Koreans, and Japanese were showing how Apple's hold could be loosened on the iPhone empire.

The celebration would be short-lived. A few months later, Google announced how it planned to comply with the Korean law: It would allow third-party payment systems to appear alongside its own mechanism. But to the surprise of Korean officials, the company said it would continue to charge app developers a commission even if they were using a third-party payment system. The commission would be slightly reduced from what they would pay using Google's in-app payment method. The 30 percent fee would be reduced to 26 percent for using outside alternatives. The 4 percent savings was essentially what a developer might get charged for a typical credit fee. In essence, Google was making it so there wasn't much of a choice.

Apple didn't make any changes, telling Korean regulators that its in-app payment system was in compliance with the new law. While the law had been hailed as allowing third-party app payment systems, the wording left some apparent wiggle room in the eyes of some lawyers. It focused on "unreasonable fees," and Apple was apparently using that as a loophole. Since its creation, Apple had argued the App Store's fees were reasonable.[2]

The collective message to the world was that Apple and Google were going to fight tooth and nail to protect their Walled Gardens. Sweeney's excitement shifted. Soon, he was on an airplane to Seoul to help rally support. "Google is disrespecting Korean law by charging a 26% fee for payments they do not process," Sweeney said during a speech before lawmakers and industry executives. "Now, it's perfectly normal for companies to charge a fee for a service it provides, but what kind of company can charge a fee for a service it does not provide? Only a monopoly can do that." Then Sweeney turned his attention to the iPhone empire. "Apple is simply ignoring Korean law," he said, before turning to the company's practices in China. "Apple complies with oppressive foreign laws which surveil users and deprive them of political rights, but Apple is ignoring laws passed by Korea's Democracy."

*   *   *

Soon it was clear that Apple's response in Korea wasn't an anomaly. After racking up fines for missing its deadline in the Netherlands, Apple in February announced a scheme similar to what Google had done in Korea. It would allow Match and other dating apps to use an alternative app payment system, but they would still be required to share 27 percent of revenue—a three percentage point savings for not using Apple's system.[3] The Dutch responded with additional fines, an extra $5.6 million, calling Apple's proposal "unreasonable" and saying it would "create an additional barrier."[4] It was the fourth fine for Apple over failing to comply with the order, totaling €20 million, not even a rounding error for a $3 trillion company.

At Apple, Phil Schiller had been working on how Apple would eventually concede to the Dutch demands. It was through both a pricing scheme that made it unattractive for developers to switch and continuing the work begun months earlier on Project Michigan to make the user experience ugly, too. By early 2022, work on Apple's Project

Michigan had been paused as the company appealed Judge Gonzalez Rogers's order to allow links out of apps. But the team was tasked with addressing similar issues in the Netherlands and those agreed to settle the Japanese investigation.

Ahead of a meeting with CEO Tim Cook, Schiller reviewed the in-app messaging for linking out. "This is not good," he told the group. They were thinking too small. "This is a big warning that the user is about to be sent out of the app to a website," he told them. "I do not think the headline should say 'continue.' This is a warning that the user is about to go out to the web and are they sure they want do that, we cannot verify that anything that occurs on the web is private and secure. The default button should not be continue." One of the team members came back with a message that she described as having more of a warning tone. A few weeks later, Schiller held a meeting for another review. Afterward, the team debriefed with Rafael Onak, one of the UX managers, about the warning messages' language. "For some reason saying, 'This app is about to take you,' feels safe to me, like don't worry, you're still within the app, and it's just guiding you somewhere else right now," Onak told the group. He saw his job as being a representative of the users, wanting to make them feel educated and well informed.

"Yeah, good point," his colleague responded. "'About to go' is a little more like into the great wide open." Instead, Onak suggested "this app is about to send" rather than "about to take" so it "doesn't sound like the app will be holding your hand." As they continued, Onak wanted something even stronger and more direct, suggesting "This app is about to send you out to an external website. You will no longer be transacting with Apple." As he thought about it more, he told the team he liked the term "out."

"If we want to scare users a bit," he wrote, "I like the addition of 'out' because it raises questions and hesitancy, haha, out, out where? OMG what do I do?"

The gamesmanship wasn't going unnoticed by lawmakers in Brussels who had been working on their own response to Apple's iPhone empire. Margrethe Vestager, Europe's top antitrust enforcer, was deep in work with European Union lawmakers crafting their response to

U.S. Big Tech, or, as she was calling them, gatekeepers. "Some gate-keepers may be tempted to play for time or try to circumvent the rules," she told an audience at the University of California, Berkeley's law school that February. "Apple's conduct in the Netherlands these days may be an example. As we understand it, Apple essentially prefers paying periodic fines, rather than comply."[5] In other words, Europe's efforts would need to be tough enough to make Apple submit.

*   *   *

Time was on Margrethe Vestager's mind. In 2019, she was appointed to an unprecedented second term as the European Commission's top antitrust officer. She had been seen by some in Europe as the commission's next president or, at the very least, a more senior role that left the day-to-day of antitrust work behind. Instead, President Ursula von der Leyen announced her reappointment during a press conference that drew audible gasps from staff and reporters.[6] In doing so, she made Vestager one of three executive vice presidents and expanded her duties to include policymaking around the digital economy. As the EU president said, she saw the competition "closely linked to the digital sector." She might have been tipping her hand that targeting Big Tech was primary to their agenda. By then, Vestager was known in the United States as the "tax lady," a nickname President Trump had given her for her pursuit of Apple's $14.5 billion in back taxes she said the company owed in Ireland.

The second term would prove challenging. Vestager's husband and youngest child had moved back to Denmark, she told a local reporter.[7] Her daughter needed a change. That left her in Brussels alone to be consumed with work when the pandemic hit. She faced a new rival with the appointment of Thierry Breton, a French businessman, as the EU's internal market chief. Whereas she was poised and calculat-ing, he was prone to Twitter skirmishes with Elon Musk. In this new term, her main focus wasn't so much combating Tim Cook as creating new tools to lay siege to Apple's domineering practices. The commis-sion had grown frustrated with the speed with which it could bring about change in dealing with fast-moving tech companies. The tax case against Apple, for example, was still in appeal. In the summer of 2020, Vestager was hit with a bruising rebuke. The European Union's second-highest court sided with Apple, saying it was annulling the decision because the commission—herself really—had failed to meet the legal standards showing Apple was improperly given special tax breaks. Vestager vowed to appeal, a process that was sure to take several more years.[8]

At that point, though, she was already working on signature pieces of legislation that held the possibility of being her legacy. One was aimed at reining in social media companies. The other was targeting the likes of Apple or, as regulators around the world were increasingly seeing its iPhone empire, a digital market. That December, she gathered reporters at the commission's headquarters at the Berlaymont for news. There, the commission announced the proposed Digital Markets Act that handed regulators stronger powers to go after companies that flout the EU's rules, including proposing fines of up to 10 percent of *global* revenue. Unlike in Washington, where Congress drafts the laws, the commission proposes the broad-brush strokes of measures, then allows the EU Parliament to weigh in before passing. Generally, the idea is that if the commission was proposing something, its members believed it could pass it—eventually, though the process could likely take years.

The announcement set Apple's lobbying efforts in Brussels into overdrive. For years, Apple had spent very little on such efforts. In 2014, for example, it spent less than €1 million. In the one-year period beginning in October 2020, the company roughly doubled its political spending to around €7 million from about €3.7 million during the previous period, records showed. It declared about seven full-time lobbyists, an increase from four and a half.[9] Maija Corinti Salvén, head of Nordic government affairs for Apple, reached out to ministers. "We at Apple fully agree that the objective of the EU legislation, and the Digital Markets Act, should be to promote competition and innovation in the EU, as well as consumer choice and protection, and—last but not least—data sovereignty," she wrote in April 2021 to Estonian officials.[10] But, she warned, "the DMA proposal in its current form, unfortunately, equates 'size' with 'harm,' and applies a one-size fits all approach to very diverse products, markets and companies." She warned that consumers would lose choice and protections—something Apple had been warning lawmakers around the world.

Publicly, Apple leaned heavily into warnings about users' privacy and security being at risk. At a large tech conference in France that summer, Craig Federighi, Apple's chief software engineer, addressed the proposed DMA. "There's a lot in the DMA that we can all get behind: doing right by users, promoting competition, and making sure that consumers have choice." But when it came to side loading—or allowing users to download apps outside Apple's App Store, beyond its rules and commission—he warned of "a step backwards in our privacy and security."

As Apple worked to water down the legislation, insiders say the broader measure was facing little resistance among the member countries, a sign of the big tech backlash that had swept Europe. Of particular importance to Vestager's team was ensuring that the anti-trust regulators could move quickly with their new tools. A key part of antitrust cases around the world dealt with defining a market.

In a California courtroom, Apple and Epic fought over what market they were talking about. Apple said video games, while Epic narrowed in on the App Store for iPhones. Such battles were often complex, time-consuming, and costly. What the DMA proposed was to streamline the process for so-called gatekeepers who met certain size requirements, allowing regulators to skip the market-definition fights and move straight to whether a gatekeeper was violating its rules. Gatekeepers would be limited in how they could privilege their own products. Side loading and third-party apps would be allowed. And, amid Apple's efforts to wiggle out of enforcement efforts in the Netherlands and Korea, the final legislation dramatically increased the penalties for failing to comply. The original proposal sought fines as high as 10 percent of a company's global revenue. That provision remained along with new fines reaching as high as 20 percent for continued violations. That summer, the European Parliament gave final approval to the DMA. It would be months before the new rules would take effect, but change was coming.

18
RETURN OF SWEENEY

2023

Apple's annual Worldwide Developers Conference looked different after the pandemic. Many of the company's biggest events—such as revealing the new iPhone—had easily transitioned to online events, attracting tens of millions of viewers on YouTube and Apple's website. The developers conference, however, had always been about bringing developers together. Part training, part marketing, part religious revival. By the spring of 2023, Apple had found a formula that included hosting hundreds of developers at Apple Park while showing them and those not on the company grounds slickly made promotional videos. The buzz this year was at a fever pitch. After years of speculation, Apple was expected to reveal its first major product since the iPhone, the Apple Vision Pro. CEO Tim Cook's vision for the future of personal computing: augmented reality. There had been other products since the iPhone's debut in 2007—the iPad, the Watch—but none that carried the same kind of weight in the minds of investors and consumers. It held the promise of changing the way users interacted with the digital world. The headset took on even greater importance as it was becoming increasingly clear that Apple's ambitions for developing an electric, driverless car to compete against Tesla had stalled out (it would eventually be axed altogether). And now Apple was ready to show its big bet, a gamble that Cook was hoping would keep the company and its App Store economy at the heart of users' lives for a generation to come. "This is a day that's been years in the making," Cook said in the video being shown to the conference and world at large. "Blending digital content with the real world can unlock experiences like nothing we've ever seen before." Investor enthusiasm would send the company's stock to new heights, closing at a price in the following weeks that would value Apple at more than $3 trillion. That marked the first publicly traded company to reach such a feat.

The headset on the outside looked like ski goggles. Inside the goggles, the user saw a screen displaying the world around them overlaid with digital images. That digital world could project a massive movie screen or unlock a three-dimensional video game. Part of what motivated Tim Sweeney to go to war with Apple was fear that Cook was going to try to take the rules of the iPhone empire into that next frontier. Sweeney fretted that the economics of paying 30 percent to Apple would carry on forever in the virtual world, or metaverse as he called it.

On that June day, Apple showed a vision that suggested Sweeney was right. As Apple revealed how the headset would work, it was very clear that the App Store was at the heart of things. "When you put on your Apple Vision Pro, you see the world and everything in it," the company announcer said. "Your favorite apps live right in front of you, but now they're in your space."

*   *   *

Atop the Hotel Zelos, a hip spot near Market Street in downtown San Francisco, Judge Yvonne Gonzalez Rogers sat on a conference stage with her counterpart on the federal bench Judge James Donato. It was an interesting pairing before a group of lawyers gathered for an afternoon of discussion about antitrust law. The federal judges were the main draw. Gonzalez Rogers had just come off the Epic Games–Apple case in which she had ordered the largest change to the App Store since its early days—a ruling that was under appeal and one that didn't go far enough for Epic's Tim Sweeney, who was now facing a $74 million legal bill from Apple. Donato's courtroom in San Francisco represented Sweeney's next best shot at bringing down the Walled Garden. While his lawsuit against Apple was rushed to the court, the twin complaint against Google and its Apple-like app store had wound its way through the federal court system. In coming weeks, it was set to be heard in the nearby Phillip Burton Federal Building in San Francisco.

Both Gonzalez Rogers and Donato were experienced enough jurists to avoid discussing specifics about their trials, especially as the Epic-Google trial was about to begin. But they talked in generalities about the state of antitrust law. It was an audience of corporate lawyers who made their careers defending the likes of Google and Apple, and those who saw the law as a way to offset the raw power of big business. To those antagonizing big business, monopoly laws were about checking power. And, in doing so, helping ensure innovation. To those lawyers, Donato had a message: He didn't think lawyers going after suspected monopolists were seeking *enough* money for the damages they had created. He subscribed to a theory that antitrust law is like the civil rights law of the economy—a tool for ensuring a level playing field for all. From his seat, he said the government was bringing little enforcement action, meaning it was civil litigation that was working to keep things in check. Some offending companies, in his opinion, were calculating that the cost of losing a lawsuit was peanuts compared with the money they could be making by breaking the law. "To me, they have clearly concluded as a business proposition it is better to break the law . . . and conspiring on prices because even if we get tagged, we're going to pay back less than what we have made through anti-competitive conduct," he told the group. "I think that's a real problem." His solution was for plaintiffs—he didn't name anyone in particular—to be more aggressive in seeking damages, including breaking up companies. "Don't be afraid to swing for the fences," the judge said.

Donato's thinking was important to know. A Google loss wouldn't affect Apple's legal case, but it had the power to sway the court of public opinion. If Google's Walled Garden was seen as illegal, why would it be okay for Apple to do something similar? Yes, the facts of the two cases were very different, and the business models were, too.

But the public's take on right and wrong was much different from that of a judge seasoned in the nuances of monopoly law. Apple might have benefited from that seasoned legal world. Google would be different. In the Apple case, Gonzalez Rogers decided on both the merits and the remedies. A jury was going to decide the Epic-Google case. And Donato would decide the remedies—if needed.

A jury trial was a much different thing in antitrust law, which had grown incredibly complex—almost like a death penalty case in Gonzalez Rogers's estimations. For a bench trial, the judge was deeply versed in the case law. Gonzalez Rogers had lots of thoughts about one of the key questions in the case, how the market was being defined. Did Apple hold an improper monopoly over accessing the iPhone marketplace, as Epic contended? Or was the market broader, such as all of the places where video games could be played? She didn't agree with their companies' arguments. With a jury, though, the companies had to tell a story—a believable one that men and women off the street could understand. "The plaintiff is the person who's humanizing for the jury," she said. To Donato, antitrust cases belonged before a jury. "This is who should be deciding antitrust cases," he said. "That's who decides civil rights. That's who decides civil rights for the economy; it should be a jury, not us—not one judge sitting by him or herself."

The *people* would soon decide.

*   *   *

A crowd had turned out to courtroom 11 on the nineteenth floor of the Phillip Burton Federal Building that Monday in early November. While pandemic precautions had relegated Epic's courtroom fight against Apple more than two years earlier to a strange setting, things had returned to mostly normal. Tim Sweeney sat behind his lawyers on one side, while Google's team sat on the other side. The wooden benches were filled, including with local law students who had been sent by their professor to witness history. The facts of the Google case would be much different from those in the Apple case, but Epic continued with its definition of the market: Google was foreclosing competition in the distribution of apps and forcing its in-app payment system. Not surprisingly, Google defined the relevant market differently, as not being limited to just Android but also including Apple and other platforms where developers can conduct digital business. What Google was going after, it argued, was taking steps to compete against Apple. And, like Apple, its lawyers said, Epic was simply violating its Play store contract in an effort to line its own pockets.

Like with the Apple case, Epic's lawyer began trying to show how the mobile phone world was different from computers where users could download software outside Google's reach. But the tone quickly changed—a sign of how differently Apple went about building its business model versus how Google did in following the iPhone's footsteps. Gary Bornstein, Epic's outside lawyer, said Google's strategy relied on two tactics to defend its digital realm: "Bribe or block." Bornstein was signaling that Google's Project Hug and Project Banyan were going to be on trial as well—all of those efforts to keep Epic's rivals and smartphone makers from breaking ranks.

"You will see this again and again, Google pays potential and actual competitors not to compete, literally gives them money and other things of value," Bornstein said. "It's like if this were a race. If competition were a road race or a foot race, it's like Google saying, 'Here's $360 million'— and that's an actual number you'll hear about—'Here's $360 million, why don't you sit this one out?' . . . That's bribe."

If it wasn't bribing companies, Bornstein continued, it was blocking them from competing by making it so hard to go around the Play store that they effectively had no choice but to use it. "Again," he told the jury, "if competition were like a race, it's like Google gets to run on a nice, smooth track and everyone else is forced to run in quicksand."

Why, this matter could be understood in ice cream, he told the jury. A town with several ice cream shops results in customers benefiting from choices in ice cream that is tastier and less expensive. A town with only one ice cream shop results in prices going up while service and quality go down. "Big companies, of course, they're allowed to compete," he told the jury. "If they can do the best and they can win that way, like, great. Good for them. But they are not allowed to use their power to stop everybody else from trying. The law draws a line between fair competition and unfair competition, between what you're allowed to do and what violates the antitrust laws, and Google has crossed that line again and again."

When it came time for Google's opening statement, Glenn Pomerantz took to the dais. He was going to hit on a theme that his team would return to over the course of the trial's sixteen days: that Google was locked in an intense fight with Apple, that it was the underdog in a global battle with the iPhone, which was the first destination for games like *Fortnite*, and that its executives were turning to contracts to ensure its users had an experience that could rival the one Steve Jobs had created. Pomerantz held up an iPhone and an Android phone. "In his opening statement, Epic's lawyer spent almost his entire opening talking about what happens on this phone, the Android phone," Pomerantz said. "What the evidence is going to show is that the people inside of Google think a lot about what happens on this phone, the iPhone, and that's because Google competes with Apple."

Unlike Apple, Google wasn't building its software and hardware alone; it was building its Walled Garden with partnerships and contracts. He pointed to the benefits Apple had in going alone. The iPhone out of the box had apps preinstalled and was ready to go. "It's a bit trickier for Android to do the same thing because each phone manufacturer controls what apps are preloaded on its phones," he said.

Pomerantz took issue with the claim of "bribe" leveled by Epic's lawyer in reference to Project Hug, which was deployed when game makers like Activision Blizzard were considering starting their own app stores. "These agreements had nothing to do with bribes," he told the jury. "They were part of Google's efforts to get the support of app developers." Those developers had a choice of where to launch their new games and apps, and Google was simply using contracts to make sure it was receiving those programs at the same time as Apple. "They didn't ask for exclusivity," he said. "You can give it to anybody you want. All they wanted was to get them at the same time so that Android users and Play customers weren't disappointed."

As for Project Banyan, the effort to persuade Samsung to give up its own app store, Pomerantz noted that the deal never got signed. "That deal never happened," he said. "He's asking you to hold Google responsible for something it did not do."

What would follow in the weeks to come was a long slog through contract negotiations with the likes of Activision Blizzard that showed how Google deployed its contracts to keep games on its store. A running theme that developed was how Google executives seemed to be taking steps to use a chat message system that didn't save their conversations, meaning there wasn't a record available of what they were discussing when making their decisions. Increasingly, the judge grew frustrated with the lack of evidence being presented, lashing out at Google's lawyers when the jury wasn't around.

Even so, there were enough emails saved that showed deals being done. And as time went on, it became clear that Epic was intent on bringing into public details of how Spotify had abandoned it in the battle against Google. For years, the two developers had joined forces in pushing lawmakers and regulators around the world to combat the power Apple held over the App Economy. Sweeney and Spotify's general counsel Horacio Gutierrez hadn't met eye to eye on the exact mission of the Coalition for App Fairness, but they had found ways to support each other. At the core, though, was a difference of what they were fighting for: Sweeney wanted to allow third-party app stores, while Spotify was against the fees being collected by in-app purchase systems. Years earlier, Spotify had stopped allowing new customers to sign up for premium service within the Apple ecosystem to avoid those fees. But Google hadn't begun charging those fees until much more recently—meaning that Spotify hadn't seen it as a fight against Google. In early 2022, as Epic licked its wounds from the results in the Apple court verdict, Spotify surprised its partner by announcing it had forged a deal with Google to offer an alternative payment method within its app. The terms weren't disclosed. "This is a significant milestone and the first on any major app store—whether on mobile, desktop, or game consoles," Sameer Samat, Google's vice president of product management, wrote in a blog post touting the deal. "This pilot will help us to increase our understanding of whether and how user-choice billing works for users in different countries and for developers of different sizes and categories."

Spotify was intent on keeping the details from being revealed in open court, but the judge had other thoughts. "Spotify's agreement with Google was the product of years of discussions between the parties, and it was the first of its kind relating to the Google Play Store," Spotify lawyers told the court in a filing trying to keep the deal secret. "If Spotify's other payments and app distribution partners learn the terms . . . this would give those partners an unfair advantage in their own contract negotiations with Spotify." When the details were revealed, however, it was hard to imagine how that was true.

Court testimony revealed that both companies agreed to spend $50 million to advertise the benefits of Spotify on Android. But Spotify won terms that meant it would pay nothing to Google if users used the streaming company's payment system to subscribe to premium service. If they used Android, Google would collect 4 percent—not 10 percent for other media apps or 30 percent for Epic. It was the sort of side deal that Sweeney had been wary of others getting when he talked about joining forces against the likes of Apple and Google. Again and again, the fight against the powers of the App Economy had shown how easily the rebellion could be swayed.

*   *   *

Judge James Donato sat at the front of the courtroom—a place of towering granite and wood walls. For sixteen days, with a break for the Thanksgiving holiday, the jury dutifully made its way to courtroom 11 to hear Epic's case against Google. It wouldn't take the jury long to render justice. After three hours and twenty-seven minutes, the jury returned to the courtroom. A stretch of time that included lunch. On all counts, the jury agreed with Epic that Google was an illegal monopoly. Later, Donato would dole out the remedies. He ordered Google to allow developers to bring their own app stores to Android for three years. Developers must be allowed to use their own billing system outside Android. And Google was entitled to charge developers only a "reasonable fee for these services"—what the judge deemed must be based on actual costs.[1]

After three years, Sweeney had won a major victory. His vision for the mobile world was coming to be—at least for a big part of it. Asked by a reporter about the two fights, Sweeney, who had sat through both trials, had his thoughts about why he won on Google and mostly lost with Apple. "The big difference between Apple and Google is Apple didn't write anything down," he said. "And because they're a big vertically integrated monopoly, they don't do deals with developers and carriers to shut down competition, they just simply block at the technical level."[2] Still, Google's loss would put increased pressure on Apple's iPhone empire. In the court of public opinion, few might see the differences between what Google had been doing and what Apple was doing.

THE IDES OF MARCH

2024

The Four Seasons Hotel in Palo Alto provided an elegant setting for one of Margrethe Vestager's final missions as Europe's top antitrust officer. She was speaking to a conference of regulators from around the world, including Japan and the U.K., along with Silicon Valley's top lawyers. Both the U.K. and Japan were moving toward their own version of the European Union's Digital Markets Act, which, after passing in 2022, was slated to go into effect that March 2024. On that date, a switch would flip for large tech companies, including Apple, that had been designated so-called gatekeepers. Under the law, Apple would need to open up its platform to other developers; what that meant exactly was still being hashed out, but if Apple failed to make regulators happy, it faced massive fines—up to 10 percent of global revenue and 20 percent if it continued to be found in violation.

Her trip was more than just speaking at a conference and enjoying the Bay Area's temperate January weather. Afterward, she would privately visit Tim Cook at Apple Park. As a general rule, Vestager didn't go to companies she was enforcing. Executives came to her. But this was something of a victory lap, something of a come-to-Jesus moment for these U.S. companies that had scoffed at the suggestion that technocrats in Brussels could try to force their rules on them. Now Europe was armed with a powerful new weapon—the Digital Markets Act—and Vestager was ready to use it in her final months in office. It was a test not just for the European Commission's willingness to use its new tools, but also for how quickly Vestager could get results. The obituary of her public career hadn't yet been written. But what had been bright and promising had given way to disappointing. Her influence appeared to be waning. Hopes of becoming EU president were dashed. Her effort to land a role leading the European Investment Bank fell short when France reportedly blocked her. She had ridden her battle with Apple and its Irish tax deal to the apex of Brussels celebrity. But as Apple chipped away at that ruling with appeals, things weren't looking good for her legacy; the only hope was the European Union's equivalent of the U.S. Supreme Court. Her antitrust case against Apple spurred by Spotify's complaint had been dragging on for years, too. The slow progress, in part, inspired the DMA's creation, giving her and her colleagues a better way of curbing big business. She attributed the relatively easy passage of the law to a collective awakening to the dangers of Big Tech.

Onstage before her foreign peers and top lawyers, she described how a new day was dawning. A revolution had occurred—even if not everyone was ready yet to acknowledge it. She drew on the classic American story of Rip Van Winkle, a lazy villager who drinks strong liquor that puts him into a deep, deep sleep, only to awake years later to discover the American Revolution has taken place and his world had dramatically changed. It was an allegory for how people can miss the radical change occurring around them in society—even if it is obvious in retrospect. In Vestager's telling, Van Winkle was a European antitrust regulator who arrived in San Francisco years earlier for the announcement of the iPhone, only to fall asleep and miss the event. He awakes sixteen and a half years later to a new world, one of digital markets held by a small number of very large companies. But he is relieved to discover that he is now armed with a new tool, the DMA, to combat abuse. And, Vestager says, he learns that the European Commission "will not hesitate to open infringement proceedings against gatekeepers that are not compliant."

A reckoning was coming. It was a message she would travel down the road to Cupertino to deliver personally to Cook, even if Apple wasn't fully convinced the changes occurring to its iPhone empire were real. In many ways, Vestager arrived at Apple Park the next day to find Apple now playing the role of Rip Van Winkle—drunk not on alcohol but, maybe, on power and, definitely, asleep to the cultural changes reshaping the world around it. As she walked into the headquarters, she thought to herself how the coldness of the place reflected the company's culture. Sterile. Without compassion. If she thought Apple would try to wiggle out of the DMA, she wouldn't be wrong. In the following days, in fact, she would watch as Apple tried to wiggle out of U.S. efforts to bring change to the App Store that were much less intense.

*   *   *

Since 2021, the rulings in the Epic Games and Apple trial had been contested. Both sides were unhappy with Judge Yvonne Gonzalez Rogers's decision that mostly sided with Apple and against Epic's argument that it held a monopoly over app distribution on the iPhone and through its demands that developers use its in-app payment system. But the judge, clearly seeing some error in Apple's actions, found fault with its practice of banning developers from being able to steer users to alternative sites for payment options outside Apple's in-app payment system.

Both Apple and Epic had appealed her rulings, which were upheld in 2023 by the U.S. Court of Appeals for the Ninth Circuit. "There is a lively and important debate about the role played in our economy and democracy by online transaction platforms with market power," the ruling concluded. "Our job as a federal Court of Appeals, however, is not to resolve that debate—nor could we even attempt to do so. Instead, in this decision, we faithfully applied existing precedent to the facts as the parties developed them." Both sides then sought relief from the Supreme Court, which declined to hear their petitions in early 2024, just days after Vestager's trip to Silicon Valley. That meant Gonzalez Rogers's ruling stood; Apple would have to make a mean-

ingful change to how it operated its App Store in the United States—
what might have initially seemed trivial compared with what Epic was
asking in total, but a change that still threatened potentially billions of
dollars' worth of sales for Apple if developers were successful in getting
users to pay for services outside the app experience.

Yet Tim Cook and Phil Schiller, the so-called App Store chairman,
seemed intent on keeping things as much in the status quo as possible. Apple's experiences in the Netherlands, Japan, and South Korea
informed their efforts. Work to comply with the U.S. order had begun
in earnest in the spring of 2023 after the appeals court loss. Project
Michigan, now renamed Project Wisconsin, was well underway in
case the Supreme Court effort failed. Working with Schiller, the team
created options for Cook on how to proceed that boiled down to two
flavors: collecting no commission for developers that steered users
elsewhere to pay, and collecting a commission that was so high it
would effectively dissuade developers from linking out.

In a June 1, 2023, meeting that included Cook and Schiller, the
executives were given a presentation that spelled out how charging
a higher commission for link outs would likely mean few developers
would go that path because they wouldn't find it "economically viable."
Developers would naturally face costs associated with providing their
own payment system, including incurring processing fees. Apple estimated the external costs would range from 5.5 percent to 12.2 percent.
Or as one of the presentation's slides said, "If we decide and had the
ability to charge a commission, we believe there would be very little
developer adoption of link-out assuming a scenario where we would
give the cost of payment discount at 3%."

But was charging a commission allowed under the judge's order?
That was the debate internally. There was disagreement if the judge
would go for it or if it was even a good look for Apple. Schiller, who
had helped guide the App Store from the beginning, was concerned.
He worried a commission could run afoul of the judge's order. Out
of everyone in the room, he was the only one who had sat across the
courtroom for all of those days in the spring of 2021, watching her
process Apple's case in real time, listening to her shred faulty arguments. "I am not on team commission/fee," he told the team in an

email late that June. "I have already explained my many issues with the commission concept."

Another concern being shared was the perception that charging a commission on links to the web might be seen as a massive overreach. "This might be perceived like we're trying to charge what happens on the internet," notes taken during one of the Project Wisconsin meetings said.

Such concerns, however, could seemingly be pushed aside to a bigger concern: the potential of losing billions of dollars if developers moved en masse to sending users to their websites rather than using Apple's payment system. Or as Oliver Carson, an App Store senior director, would later say, "All else being equal, the lower the commission, the greater financial hit to Apple."

If Schiller was Team No Commission, Chief Financial Officer Luca Maestri and his finance cohorts were squarely Team Commission. And the question for them, in meetings, centered on how much they could get away with. For a while, they debated 20 percent or 27 percent. The 20 percent proposal was problematic for several reasons. If they went that route, the Project Wisconsin team was struggling with justifications for why Apple had 20 percent in the United States while it collected 26 percent in Korea.* Very quickly the idea of 20 percent as a commission fee was abandoned. Maestri and Cook wanted 27 percent. Avoiding no commission and managing to institute a 27 percent commission for just a year or eighteen months would save Apple hundreds of millions if not billions of dollars.

If small and midsized developers saw zero value in abandoning Apple's payment system for the three percentage point discount, Apple still had to worry about large developers who might have some economies of scale. Years earlier, Apple had lowered its fees to 15 percent from 30 percent for subscription models around apps like HBO and *The New York Times*. Executives decided that in order for those apps to remain eligible for that pricing, they had to keep using

---

* While Apple had initially not offered an alternative in South Korea to its in-app payment system, it eventually relented. In 2022, it offered the option to developers to use a third-party payment system while still charging a commission slightly less than 30 percent.

Apple's payment system. They could steer customers to external payment options, but then their own in-app experience would see costs skyrocket.

And then there were the psychological effects that Apple wanted to impose on app users in hopes of deterring them from using those in-app links. The judge's orders prohibited Apple from banning "buttons, external links or other calls to action that direct customers to purchasing mechanisms." How that was interpreted was of great concern inside Apple. Initially, according to meeting notes, Project Wisconsin figured that if Apple charged a commission on links out, developers would be given essentially free rein on the design of those buttons and links. But if there was no commission, they expected to curtail the look and placement of the buttons and links. That's not the way things worked out, however.

Apple would allow developers to provide a link out but not present the alternative payment option to users at the same time or place as the in-app payment system. So, for example, if the app had a section for shopping, an external link could be nowhere near that screen. "How much can we limit what devs do with the text and links?" one member of the Project Wisconsin team asked by email. They also insisted on warning screens if users clicked on the link and, *if* they continued, users would be taken to a static website where they would have to reenter their username and password—not a page where they were already logged in and ready to make the transaction. In other words, the more complicated things became for the user, the more likely they wouldn't follow through with the sale. "We know it's very likely that when a link-out happens, there will be some breakage, meaning customer [*sic*] dropping off during the buy-flow process due to a less seamless experience compared to Apple's IAP," according to notes of a presentation given to Cook in June.

Furthermore, the presentation said, there would be a tipping point for the developer between where they lose more money on linking out than they make sticking with Apple's commission. And they knew the addition of scare screens and the need to sign into accounts would create additional friction and mean more breakage. Months earlier, Apple had spent time trying to make its link-out messages in the Netherlands and elsewhere as dire as possible. But in a June meeting, Cook told the team to revise the warnings further "to reference the fact that Apple's privacy and security standards do not apply to purchases made on the web."

Following the Supreme Court decision, Apple was ready for its response to Gonzalez Rogers's order. Apple rolled out its new rules for steering customers out of apps—including announcing it would take 27 percent of spending outside the app made seven days after the user clicked on the link (or 12 percent if a developer was part of its small business program).[1] It also detailed how developers could provide outside links. Predictably, Epic and many others were unhappy. "Apple filed a bad-faith 'compliance' plan for the District Court's injunction," Sweeney posted on social media. "Apple has introduced an anticompetitive new 27% tax on web purchases," he added. "Apple has never done this before, and it kills price competition. Developers can't offer digital items more cheaply on the web after paying a third-party payment processor 3–6% and paying this new 27% Apple Tax." Very soon his lawyers were filing a complaint with Gonzalez Rogers. And she'd act quickly to review Apple's actions, holding several days of hearings, including pulling Schiller back onto the stand to explain the company's actions. It became pretty clear that the judge wasn't happy with Apple. At one point, she took issue with Apple's demands that the link-out buttons look a certain way, which Epic's lawyers suggested would deter users from actually using them. Matt Fischer, head of the App Store, denied that was the case. "I can't imagine a logical reason why Apple would demand that of competitor apps," the judge interrupted. "What's the logical objective reason for, not suggesting it, but demanding it?"

Fischer floundered with an answer about it being "early days."

She wasn't buying it. "You still haven't answered my question. Other than to . . . to stifle competition . . . I see no other answer. Give me one."

Fischer tried again. "This was the decision that was made . . ."

"That's not my question," she said, cutting him off. She continued, "When you were all talking about this, was any rationale provided for requiring this, not suggesting it, but requiring it when you all know that what it will do is stifle competition? Any other reason given? Because if you cannot identify one, that's my assumption."

She had him. She had Apple. "I don't remember exactly what this was . . . in the discussion around the . . . the button styles, Your Honor."

"So," she concluded, "the answer is no. You cannot think of any other reason for requiring that."

"No," he responded.

As members of Apple's team listened to the judge question their actions, they vented among themselves. "It's our FUCKING STORE," Marni Goldberg, a public relations director, texted a colleague at one point during the hearing. Goldberg, a seasoned PR pro, knew it was bad. The judge's words would live on beyond the court's walls. "This will be a quote for sure in coverage," she texted to her colleague Hannah Smith. "'Other than to stifle competition.'" She fumed further that Fischer was their weak link and Epic's lawyers had spotted it. "I wonder if they're spending so much time here because they know Matt is the worst of our witnesses."

Smith shot back, "I think your guess is accurate. They knew we didn't plan to call him." As the hearings continued, Goldberg kept dwelling on Fischer's performance. After another Apple executive struggled on the stand, Goldberg texted that she hoped he didn't get fired but was still stewing on Fischer. "Maybe Matt deserves to."

A few months later, in fact, Fischer retired from the company after more than two decades. It would be many more months before Epic's issues would be resolved because Apple would apparently drag its feet on disclosing documents in the case.[2] Many more months of costly lawyer fees, many more months of Epic not being in the App Store in the United States, many more months of uncertainty.

*   *   *

The promise of Margrethe Vestager's Digital Markets Act was to avoid endless foot-dragging that benefited the deep pockets of global corporations with armies of lawyers, corporations that profited from delays as they continued doing what regulators were trying to stop. Shortly after rolling out the 27 percent fee for link outs in the United States,

Apple introduced rules in Europe—only for the EU bloc—in an effort to comply with the DMA. It included the link-out commission of 27 percent. For in-app purchases, it lowered its in-app fee to 17 percent from 30 percent. Those who use the in-app payment system would be charged an additional 3 percent. Apple also introduced a new "core technology fee" that charges developers 50 euro cents for every download of their apps after they reached one million downloads in total or within twelve months. That per-download fee would apply to free apps (not government, education, or nonprofits) and regardless of whether the download occurred within the App Store or outside it. The biggest change was announcing it would allow apps to be downloaded outside its App Store, though Apple included new rules that gave it some control over those programs. "The changes we're announcing today comply with the Digital Markets Act's requirements in the European Union, while helping to protect E.U. users from the unavoidable increased privacy and security threats this regulation brings," Phil Schiller said in a statement to the press. In response, Tim Sweeney called Apple's plan "hot garbage" and "a devious new instance of Malicious Compliance." It didn't take long for Vestager to agree. In June, she charged Apple with violating the DMA, marking the first time European regulators used the new powers vested in them. In particular, the commission was unhappy with the "core technology fee" and other requirements as well as the commissions charged on developers who steer customers to other payment options. Apple faced fines of up to 10 percent of its global revenue, which was more than $380 billion in the previous fiscal year. And if things continued, that penalty could rise to 20 percent. Or in other words, Apple might face a $38 billion fine or possibly $76 billion, though it would ultimately be much less. The charges were preliminary. Apple said it was confident it complied with the law. And it defended its rules and fees as being fair for what it was providing developers.[3] "The European Commission would like Apple to open its ecosystem, and Apple is saying no way," Tommaso Valletti, a former top economist at the commission, told *The New York Times*. "Apple is basically saying, 'See you in court.'"

The decision came shortly after another blow by Vestager to Apple. In March, she concluded work on the case brought by Spotify years earlier. At the heart of the case were rules by Apple that prevented Spotify from communicating with its users about ways to pay for upgrade subscriptions outside Apple's in-app payment system. "Apple's conduct, which lasted for almost 10 years, may have led many iOS users to pay significantly higher prices for music streaming subscriptions," the commission said in its blistering statement. The penalty was steep. The European Union fined Apple about $2 billion for setting unfair rules for music-streaming apps. The fine marked one of the largest ever by the bloc's antitrust regulators and was much larger than some antitrust lawyers had expected. "I think it's important to say that if you are a company who's dominant, and you do something illegal, you will be punished," Vestager said.[4] (Apple promised to appeal, saying the commission have failed to uncover any credible evidence of consumer harm.")

In short order, Vestager was landing her final blows on Apple's iPhone empire. The only remaining question was her first big fight: the Irish tax deal. The European Court of Justice, essentially the Supreme Court of the EU, answered that soon after. Delivering a surprising blow against Apple, the court upheld the commission's order for Ireland to recoup $14.5 billion in unpaid taxes from Apple, plus interest. *Politico* called it a "stunning court win." The ruling came in the final weeks of her tenure as Europe's top antitrust officer. Together, the three actions against Apple could potentially total more than $50 billion—or almost 60 percent of its profit that year. Standing before reporters, Vestager admitted the emotion of the moment. "It was a win that made me cry because it is very important," she said.[5] In private, Vestager celebrated with a glass of champagne.

*   *   *

As Apple faced losses in Europe, its empire's fate in the United States was to be further tested in an unusual way. While Google fought Tim Sweeney in a San Francisco federal court, the search giant faced government lawyers in a Washington, D.C., court over another antitrust case about the power of its search business, which answered about 90 percent of all internet search queries globally. At the heart of the case, filed in the waning days of the first Trump administration in late 2020, was Google's deal with Apple that made its search engine the default on the iPhone. It was an agreement that provided Apple with $20 billion in 2022. And it was just one of the ways that Cook had figured out how to profit off the iPhone over and over again: first from the sale of the device, then by charging as much as 30 percent on in-app purchases and again by getting a cut of users clicking on Google ads using their Apple devices. Google was on trial, but Apple might as well have been, too. Its executives would take the stand in late 2023 as they defended their actions. Eddy Cue, the senior Apple executive who had negotiated the troublesome e-book deals years earlier, was again called on to explain why his company hadn't overreached. "When we're picking search engines, we pick the best one and we let the customer easily change them," he told the judge. "So, I have no problem with that. I think we're doing the right thing by customers."

An important part of the Justice Department's case against Google revolved around how, the lawyers argued, the Apple deal hindered Microsoft from competing in the search space. Microsoft's chief executive, Satya Nadella, told the judge from the witness stand that the deal hindered his company's ability to compete in the market, threatening rivals from being able to crack the quickly developing space of artificial intelligence. Breakthroughs in AI were upending Silicon Valley because the new technology held the potential—and threat—of unseating established players just as the iPhone had done. But it wasn't yet clear how. Nadella told the judge that Google had a distribution advantage with its default deal. "The distribution advantage Google has today doesn't go away," Nadella said. "In fact, if anything, I worry a lot that— even in spite of my enthusiasm that there is a new angle with AI—this vicious cycle that I'm trapped in could even become even more vicious because the defaults get reinforced."

During the trial, the judge seemed to agree the Google-Apple relationship was troubling, saying he didn't see how a new entry could entice Apple away from its partner. "I can't conceive of a world in which some other competitor, particularly a new competitor, could do that if Microsoft couldn't do it," the judge said.

For its part, Google argued modern internet users have many ways to find information and shop online, including Amazon.com and TikTok. The company said it competed fairly and became the market leader because it provided the best search engine. Lawyers often noted that a top search query on Microsoft's Bing was "Google."[6] In the end, however, the judge sided with the government, ruling Google was in the wrong. "Google is a monopolist, and it has acted as one to maintain its monopoly," Judge Amit Mehta wrote in his 276-page decision. He mentioned the word "Apple" almost three hundred times in his decision. "Apple, a fierce potential competitor, remains on the sidelines due to the large revenue share payments it receives from Google."

As part of the remedies, the government proposed that Google be banned from doing the kinds of deals that were paying Apple so much money. The Justice Department wasn't done with Apple either.

*   *   *

Almost a decade earlier, Spotify had hired Jonathan Kanter to help it make the case to the U.S. government that Apple was a monopolist that needed to be held to account. It was ahead of its time. But the work that Spotify, Kanter, and others would do in the years to follow would lay the groundwork for the once inconceivable: a landmark claim by the Justice Department that Apple's iPhone empire had grown too powerful.

While work for a case against Apple had begun in the Trump administration, it took on new urgency with Kanter's appointment as the Justice Department's chief antitrust officer in 2021 by the Biden administration. Given his work over the years on the other side of Apple and Google cases, some in Washington argued he should recuse himself from those cases. But in private and public practice, he had always been on the side of trying to rein in those giants. His loyalty to Apple's enemies had helped make him a rich man. He split with his law firm Paul, Weiss in 2020 when it brought on longtime Apple lawyers from a rival firm. Instead of giving up the likes of Yelp, he took a $20 million buyout and set up his own firm. His work against Big Tech won him allies, such as Senator Elizabeth Warren, who backed his appointment.[7] And now, heading the eight-hundred-person antitrust division, Kanter made it clear internally and externally that he was going big-game hunting. In an interview with *Politico*, he said he was on a campaign to "reinvigorate antitrust enforcement."

His team had the benefits—and challenges—of having watched Epic Games' case against Apple. It was something of a split decision. Mostly it went Apple's way, except for the judge's ruling that Apple must stop banning developers from steering users to other ways of paying for in-app purchases. Epic's argument that Apple held an improper monopoly over distributing apps and requiring its in-app payment system had fizzled. Still, the judge had suggested there was still room for another case—with different arguments—to find Apple was acting improperly. "The evidence does suggest that Apple is near the precipice of substantial market power, or monopoly power, with its considerable market share," she wrote.

Kanter agreed. To him, Apple's conduct was classic monopolist behavior. Unlike others who were arguing for new laws to hold the likes of Apple accountable, he saw a case with the current law, a case similar to what had been brought against Microsoft. Years earlier, Kanter had argued Spotify's case to the government that a Microsoft-like argument against Apple was possible. But that was before Apple had grown even bigger and into new areas. As the Justice Department lawyers began working up their case, they zeroed in on a broader case against Apple than what Epic had brought and what Spotify had done in Europe. Whereas plaintiffs in antitrust cases often go narrow, Kanter was going big: smartphones. As his team defined it, Apple held more than 70 percent of so-called performance smartphones in the United States and more than 65 percent of the broader smartphone market. But beyond that, the team argued, Apple held dominance over young people with one-third of all iPhone users in the United States being born after 1996 while just 10 percent of Samsung buyers were that young. The team was focused not just on the App Store but on all of the ways in which Apple held sway over the iPhone empire, arguing it had a duty to allow rival software and hardware makers to work more seamlessly with its product. Apple held sway over how smartwatches connected to the iPhone, how streaming gaming interacted with the device, how digital banking worked, how users texted each other with blue or green bubbles.

As they prepared a complaint against Apple, Microsoft was clearly on their minds—a constant comparison, mentioned twenty-six times throughout the eventual filing. "In 1998," the government lawyers wrote, "Apple co-founder Steve Jobs criticized Microsoft's monopoly and 'dirty tactics' in operating systems to target Apple, which prompted the company 'to go to the Department of Justice' in hopes of getting Microsoft 'to play.' But even at the time, Apple did not face the same types of restrictions it imposes on third parties today."

On March 21, Kanter's office filed its antitrust complaint against Apple. The company was prepared with a full-throated promise of fighting back. "This lawsuit threatens who we are and the principles that set Apple products apart in fiercely competitive markets," Apple said in a statement distributed to the media. "If successful, it would hinder our ability to create the kind of technology people expect from Apple—where hardware, software, and services intersect."

The outcome would matter, of course. But the damage had already been done. In the eyes of many, Apple had become one of the robber barons of a new era.

EPILOGUE

2025

Tim Cook called Donald Trump. As momentum in the final days of the 2024 presidential election seemed to be building to send the former president back to the White House, the Apple CEO joined other business leaders in reaching out to pay homage. Cook had skillfully navigated Trump's first term, forging a relationship with the Republican. In that time, Trump had shown he enjoyed the personal attention of America's biggest business leaders. He liked the headlines that came with pledges of American investments. He liked the photo ops. That relationship helped Apple, during the first Trump administration, navigate a trade war with China, getting the iPhone exempted from U.S. tariffs imposed at the time on Chinese made goods.

The halo of the Apple's brand was still bright in Trump's eyes, evident by the fact that he quickly touted publicly the most recent call with Cook. Perhaps in Trump's mind it was akin to an endorsement of his candidacy. Or, at least, a way for Trump to project that powerful players were taking him seriously. And that Cook needed *his* help. On the call, according to Trump, Cook complained about all of the money being levied against Apple by the European Union—the recent tax loss and the $2 billion penalty in the Spotify case. "I even said about Apple, 'Can you pay that?' . . . That's a lot of money," Trump said recounting the conversation on a podcast a few hours after the call. "I said, 'But Tim, I got to get elected first but I'm not going to let them take advantage of our companies.'" As he talked about his relationship with Cook, Trump sounded impressed. "I believe that if Tim Cook didn't run Apple—if Steve Jobs did—it . . . wouldn't be nearly as successful," Trump said.

Such warm talk worried some antitrust observers that the political winds would shift if Trump was elected. In particular, the prospect of a second Trump term unnerved some in Brussels. Margrethe Vestager was leaving the European Commission and some in Berlaymont wondered if the new European Commission would have the political will to keep fighting Big Tech in an era of heightened transatlantic tensions. The Europeans had ridden a wave of momentum against Apple. If that slowed, it would seemingly help Apple. The longer Cook stretched out its fight with appeals, the more he increased Apple's chance that the drive against the company would lessen. The politics and politicians would change. The company, with all of its money and now more than one billion iPhone users, could outlast the antagonists. Then add Trump back into the White House, which could introduce a new hesitancy among Europe to flex the full power of the new Digital Markets Act, especially as another era of trade tensions appeared on the horizon.

While Trump had shown a willingness to tussle with Europe, his relationship with Big Tech was also complicated. It was, after all, his Justice Department that had begun an inquiry into Apple's power back in 2019. That effort had snowballed into 2024's sweeping case against Apple, albeit filed under the Biden administration. And as Trump talked about his affection for Cook, Elon Musk, the unlikely Trump surrogate, was sending his own warning to Apple.

Musk had grown increasingly active politically since acquiring Twitter, which he renamed X. The social media platform—coupled with his own personal celebrity and new willingness to spend his fortune helping get Trump elected—had made Musk a political force. He was even on the campaign trail in the battleground state of Pennsylvania in the weeks before Election Day, drawing large crowds and holding hours-long Q&As with supporters. During those events, he talked Mars, Tesla, and free speech. Asked about his vision "for challenging the existing cellphone syndicates," Musk used the moment to send Apple a message. "The idea of making a phone makes me want to die," Musk said to laughter. "But if we have to make a phone, we will." He name-checked Apple and Google's Android, amid a scattering of "boos" from the large crowd outside of Philadelphia. "They need to make sure they don't have a heavy hand in . . . the App Store,"

Musk continued. "Or they will create a forcing function for a new competitor." Like in 2022, Musk was again stirring the pot on Apple, a reminder that the company's control over the App Economy had become a political issue, too.

Less than three weeks later, Trump was reelected. Musk was soon camping out at Trump's Mar-a-Lago Club in South Florida as the incoming administration prepared to take office in January. A parade of Big Tech leaders soon descended, including Cook. Clearly, some of it was to kiss the ring. But efforts were also being made to reframe how Big Tech was seen within the Republican administration. Facebook co-founder Mark Zuckerberg, who had been a target of Trump's ire and whose company was also fighting claims of anticompetitive behavior by the Federal Trade Commission, went on Joe Rogan's podcast. There, Zuckerberg tried to paint Apple as the real threat. By his back-of-the-envelope math, Zuckerberg said, his company would have made twice as much profit without Apple's "random rules."

As the tech lords vied to recast themselves as part of the good Big Tech, they also raced to give large donations to his presidential inauguration. Among those gifts, Cook gave $1 million. On inauguration day, it seemed to pay off; Cook, along with Musk, Zuckerberg, and the others, were on stage at the Capitol, watching just feet away from Trump being sworn in. But was it a new day for Big Tech?

*   *   *

In the opening weeks of the second Trump White House, the first indication of how aggressive the new administration might be against Apple dropped in a legal filing by the Justice Department in one of its cases against Google. This was the case that involved the deal Google gave Apple to make its search engine the default on iPhones. A judge had ruled at the end of 2024 that Google was engaged in anticompetitive behavior and was holding hearings on remedies. The government wasn't backing down. The Trump administration's Justice Department was reiterating what the DOJ under President Joe Biden had sought to do: break up the search giant.

The filing came ahead of Trump's appointment to replace Jonathan

Kanter as head of the Justice Department's antitrust office. During her Senate confirmation hearing, Gail Slater, a longtime Washington lawyer, expressed worry about the dominance of online platforms. She was seen as an antitrust enforcer who would be friendly to corporate mergers but tough on tech. During her first public speech after taking office, Slater cheered the work against Google and talked about how antitrust laws needed to be viewed as being in a new era to protect individual liberty. Whereas monopiles a hundred years ago could control prices and exclude competition, she said, "today's online platforms can do so much more. They control not just the prices of their services, but the flow of our nation's commerce and communication."

By spring, the European Commission was also ready to show its cards on how much confidence it had in flexing its power from the Digital Markets Act. In early 2024, European regulators had announced Apple was in breach of the DMA for preventing app developers from freely steering users to alternative payment channels outside of their apps. It took issue with the 27 percent fee it charged for purchases made outside of the app using the links. The decision kicked off a twelve-month proceeding that led to a late April 2025 announcement that the European Union was fining Apple €500 million—not the 10 percent of global revenue that it could have levied. A warning shot. Apple, which said it would appeal, had sixty days to comply or face more penalties.

*   *   *

The power of the iPhone was the user experience. It was at the heart and soul of the product.

Tim Cook has often said as much. When he testified before Judge Yvonne Gonzalez Rogers during the sixteen-day trial in 2021 over whether Apple had an illegal monopoly over the iPhone, he defended the company's actions in large part by effectively saying everything was done for the user. "We have a maniacal focus on the user, in doing the right thing by the customer," Cook testified. "We take a lot of the complexity of technology away from the user," he added, "and make things simple, not complex."

It was a point that clearly resonated with Gonzalez Rogers, who cited the importance of consumer choice in her ruling in the case that mostly went Apple's way. She hadn't gone along with Tim Sweeney's demands that Apple allow third-party app stores. Nor had she agreed with Epic Games' pleas for alternative in-app payment methods. However, Gonzalez Rogers was troubled by one particular practice that Apple enforced: a so-called anti-steering provision within the App Store rules that meant companies, such as Epic, couldn't tell users that an alternative, perhaps cheaper, way of paying for in-app content might be available outside the Walled Garden of Apple's iPhone empire. This was anti-competitive to Gonzalez Rogers because the rules "hide critical information from consumers and illegally stifle consumer choice." She ordered it stopped. Apple disagreed and appealed—which was its right. After exhausting those avenues, Cook and his team rolled out a solution that on its face looked to make external links so un-user friendly as to make them basically useless. Or that's what Epic argued when it sought to have the judge address the matter.

The judge had originally ruled that Apple was entitled to collect a fee for the technology and services it created with the in-app payment system. She had given a light touch on how it should address her demands against the anti-steering provision. Apple wasn't allowed to stop developers from including "buttons, external links, or other calls to action that directed customers to purchasing mechanisms, in addition to In-App Purchasing."

Gonzalez Rogers spent more than a year trying to understand Apple's motivations for rolling out a steering rule that *did* allow external links yet seemed to fall short of the spirit of her order. She clearly grew frustrated by what she learned. In her view, Apple not only defied her orders, it worked to cover up its efforts and actively deceive. Court records revealed how Phil Schiller, the so-called App Store chairman, had warned his colleagues that their proposed efforts to collect a big commission on external purchases wouldn't fly with the judge. Yet they ignored him, worried about losing out on possibly billions of dollars if developers adopted the new workaround en masse.

Gonzalez Rogers's long-awaited ruling on the matter came on April 30, 2025. It was scathing, almost personal in tone, and called out Cook by name for his role. "Apple *willfully* chose not to comply with this Court's Injunction," she wrote in an eighty-page finding. "It did so with the express intent to create *new* anticompetitive barriers which would, be design and in effect, maintain a valued revenue stream; a revenue stream previously found to be anticompetitive." She called it a "gross miscalculation" by Apple to think she would tolerate such insubordination. "The cover-up made it worse," she concluded. "For this Court, there is no second bite at the apple."

Immediately, she ordered, Apple was to stop imposing any commission or fee on purchases that users made outside an app in the U.S. Whereas she had previously been broad her in instructions, this time she went very detailed. No restrictions to style, language, formatting, quantity, flow or placement of links for purchases outside the app. No limitations of buttons or calls to action. No audits. No monitors. No tracking. No reports of purchases made. She even dictated what the language of the dialogue box would say to notify users they were leaving the app: "Open in 'Safari'? You will leave the app and go to the developer's website."

Furthermore, she concluded that one of Apple's vice presidents "outright lied under oath" when he testified that Apple hadn't looked at the cost of alternative payment solutions for developers who took advantage of the links out for purchases. In fact, the evidence, showed that's exactly what Apple did. It picked a 27 percent commission fee knowing that payment solutions would "conveniently exceed the 3% discount Apple ultimately decided to provide by a safe margin," she wrote. More than that, the executive said that Apple had no idea what fee it would impose until January 2024 when, in fact, according to Apple's own internal records it had 27 percent as part of its main plan determined way back in July 2023.

The judge announced she was referring the violations to the Justice Department to investigate Apple for possible criminal contempt. And she singled out Cook for ignoring Schiller and going forward with a plan she ultimately decided wasn't good for the user but was good for Apple. She wrote: "Cook chose poorly."

For Cook's part, he said he disagreed with the findings. "We've complied with the court's order, and we're going to appeal." Nevertheless, the ramifications were immediate. Spotify, which had taken its fight for the right to link externally from Washington, D.C., to Brussels and around the world, was ready the next day. It submitted an update to its app that took advantage of the court-ordered changes. Soon, Spotify's update was approved and, the company would say, quickly saw a significant increase in users upgrading their accounts through its Apple app. "After nearly 10 years, Spotify can now show pricing + direct purchase links in our app for U.S. users," Daniel Ek, Spotify's founder and CEO, announced on social media. "This is a huge win for consumer choice and tech innovation. At Spotify, we've always believed an open internet is a much better internet—so we couldn't be happier to see this process being made." Amazon, too, quickly responded, updating its Kindle app for iPhone users, providing a link to buy digital books.

Perhaps no one was more excited than Sweeney, who spent the next few days touting the win on his social media account. "Game over for the Apple Tax," he declared in one post. He'd won ways around the in-app payment system in both the U.S. and Europe. "4 years 4 months 17 days," he noted along with a screen shot of the video he originally published when launching Project Liberty, the piece he commissioned satirizing Apple's own broadside against another era's big-tech monster, IBM. "Epic Games has defied the App Store Monopoly. . . . Join the fight to stop 2020 from becoming '1984'. #FreeFortnite." *Fortnite* would soon be back on the App Store as well, rising to No. 1 among free apps.

Sweeney had argued that as digital devices had become center to users' lives, freedom in the real world required freedom in the digital world. His belief had cost Epic a lot—more than $1 billion, by his estimate, in legal fees and lost revenue. Still, the return of *Fortnite* to the App Store would give him hope that the rest of the world would follow the path forged in the U.S. and Europe. There were signs he might be right. The U.K. had followed through with new regulations similar to the European Union's Digital Markets Act. Brazil regulators were challenging Apple's control. And, amid trade tensions with China, Apple's App Store had reportedly come under more scrutiny.

The weekend after the win, Sweeney celebrated. Not in the digital world, but on a hike at a public park near Epic's headquarters, lush with a green forest and open with possibilities—and without a Walled Garden in sight.

ACKNOWLEDGMENTS

A book like this is built on the shoulders of others who have gone before, drafting the first takes of history with their daily news coverage. Apple beat reporters, such as Tripp Mickle and Mark Gurman, spend their days trying to understand what's really going on behind the Walled Garden.

I'd like to thank former *Wall Street Journal* editors Matt Murray, Karen Pensiero, and Jamie Heller for supporting me to do this book. Heller, in particular, long pushed me to do the best work possible and taught me so much along the way. I'd also like to thank the *Journal*'s current leadership team, Emma Tucker, Liz Harris, and Charles Forelle, for supporting this project. It's an honor to work with them.

The team at HarperCollins, especially editor Sean Desmond, made this book possible and their work is most appreciated. My agent, Eric Lupfer, is the best. I'm also grateful to my research team: Ross and Drew, who provided firsthand knowledge of *Fortnite* that was especially helpful.

NOTES

CHAPTER 1: AN UNREAL HERO, 2006

1. Rus McLaughlin, "The History of Gears of War," *IGN*, Jan. 18, 2012, www.ign.com /articles/2011/09/12/the-history-of-gears-of-war.
2. Cliff Bleszinski, *Control Freak: My Epic Adventure Making Video Games* (Simon & Schuster, 2022).
3. Ibid.
4. Ibid.
5. Tom Bissell, "The Grammar of Fun," *The New Yorker*, Oct. 27, 2008, https://www.newyorker .com/magazine/2008/11/03/the-grammar-of-fun.

CHAPTER 2: "WITNESSING HISTORY," 2008

1. Leander Kahney, "Straight Dope on the IPod's Birth," *Wired*, Oct. 17, 2006, www.wired .com/2006/10/straight-dope-on-the-ipods-birth/?tw=rss.index.
2. Peter Burrows and Adam Satariano, "Can Phil Schiller Keep Apple Cool?," Bloomberg News, June 7, 2012, https://www.bloomberg.com/news/articles/2012-06-07/can-phil -schiller-keep-apple-cool?sref=PRBlrg7S.
3. Nitin Ganatra, interview by David C. Brock and Hansen Hsu, April 24, 2017, Computer History Museum Oral History, Mountain View, California, https://archive.computerhistory .org/resources/access/text/2018/05/102738249-05-01_acc.pdf.
4. Jay Rubin, interview by author, 2024.
5. Steve Jobs, interviewed by Nick Wingfield, Aug. 7, 2008, Cupertino, California, *Wall Street Journal*, https://www.wsj.com/articles/the-mobile-industrys-never-seen-anything-like-this -an-interview-with-steve-jobs-at-the-app-stores-launch-1532527201.
6. Quinn Myers, "The Inside Story of IBeer, The Underdog Beer App That Made Millions," Mel Magazine, (2022), https://melmagazine.com/en-us/story/ibeer-app-history.
7. Myers, "The Inside Story of IBeer."

CHAPTER 3: "MORAL RESPONSIBILITY," 2009

1. Ryan Singel, "Apple App Store Bans Pulitzer-Winning Satirist for Satire," *Wired*, April 15, 2010, web.archive.org/web/20100903084205/http://www.wired.com/epicenter/2010/04 /apple-bans-satire/.
2. MG Siegler, "Steve Jobs Reiterates: "Folks Who Want Porn Can Buy an Android Phone," *TechCrunch*, April 19, 2010, web.archive.org/web/20100826073009/http://techcrunch .com/2010/04/19/steve-jobs-android-porn/.

3. Chet Haase, *Androids: The Team That Built the Android Operating System* (No Starch Press, 2022), 24.

4. Ibid., 251.

5. Eric Chu, "Android Market: a user-driven content distribution system," Android Developers Blog, Aug. 28, 2008, android-developers.googleblog.com/2008/08/android-market-user-driven-content.html.

6. Eric Chu, "Android Market: Now available for users," Android Developers Blog, Oct. 22, 2008, android-developers.googleblog.com/2008/10/android-market-now-available-for-users.html.

7. Chet Haase, *Androids: The Team That Built the Android Operating System* (No Starch Press, 2022), 308.

8. MG Siegler, "Apple Rejects App For Using an Icon That Somewhat Resembles an iPhone," *TechCrunch*, April 20, 2009, techcrunch.com/2009/04/20/apple-rejects-another-app-for-using-an-icon-that-looks-like-an-iphone/.

9. MG Siegler, "Like My Parents in 1994, Apple Finds NIN's The Downward Spiral Objectionable," *TechCrunch*, May 2, 2009, techcrunch.com/2009/05/02/like-my-parents-in-1994-apple-find-nins-the-downward-spiral-objectionable/.

10. Sarah Perez, "Apple Rejects 'Politically Charged' iPhone App," *New York Times*, Sept. 29, 2009, archive.nytimes.com/www.nytimes.com/external/readwriteweb/2009/09/29/29readwriteweb-apple-rejects-politically-charged-iphone-app-9006.html.

11. Brian X. Chen, "School-Shooting iPhone Game Removed From App Store," *Wired*, July 20, 2009, www.wired.com/2009/07/zombie-school/.

12. John Gruber, "Ninjawords: iPhone Dictionary, Censored by Apple," *Daring Fireball*, Aug. 4, 2009, daringfireball.net/2009/08/ninjawords.

13. John Gruber, "Phil Schiller Responds Regarding Ninjawords and the App Store," *Daring Fireball*, Aug. 6, 2009, daringfireball.net/2009/08/phil_schiller_app_store.

14. Jenna Wortham, "Apple Gives App Developers Its Review Guidelines," *New York Times*, Sept. 9, 2010, www.nytimes.com/2010/09/10/technology/10apple.html.

15. Richard Williamson interview by Hansen Hsu and Marc Weber, Oct. 12, 2017, Computer History Museum Oral History, Mountain View, California, https://archive.computerhistory.org/resources/access/text/2018/07/102740223-05-01-acc.pdf.

16. Matthew Shaer, "Bloomsday: Apple reverses course on controversial Ulysses Seen iPad App," *Christian Science Monitor*, June 16, 2010, www.csmonitor.com/Technology/Horizons/2010/0616/Bloomsday-Apple-reverses-course-on-controversial-Ulysses-Seen-iPad-app.

CHAPTER 4: A BOOK TOO FAR, 2010

1. Fred Vogelstein, "Why Is Obama's Top Antitrust Cop Gunning for Google?," *Wired*, July 20, 2009, www.wired.com/2009/07/mf-googlopoly/.

2. Press release, "Competition: EU and US Celebrate 20 Years of Cooperation . . ." European Commission, Oct. 13, 2011, Eec.europa.eu/commission/presscorner/detail/en/ip_11_1194.

3. Thomas Catan, Jeffrey A. Trachtenberg, and Chad Bray, "U.S. Alleges E-Book Scheme," *Wall Street Journal*, April 11, 2012, www.wsj.com/articles/SB10001424052702304444604577337573054615152.

CHAPTER 5: BIG BUSINESS, 2010–2011

1. Erik Brudvig, "Undertow Review," *IGN*, May 13, 2012, www.ign.com/articles/2007/11/20/undertow-review.

2.   Mike Bowden, "Undertow Review," *Eurogamer*, Dec. 9, 2007, www.eurogamer.net /undertow-review.

3.   Neal Pollack, "Spotify Is the Coolest Music Service You Can't Use," *Wired*, Dec. 27, 2010, www.wired.com/2010/12/mf-spotify/.

4.   Anne Steele, "Spotify Dominates Audio Streaming, but Where Are the Profits?," *Wall Street Journal*, Jan. 18, 2024, www.wsj.com/business/media/spotify-streaming-music -podcasts-audiobooks-3e88180d.

5.   Glenn Peoples, "Spotify CEO: U.S. Launch is 'Looking Pretty Good,'" *Billboard*, Feb. 3, 2010, https://www.billboard.com/music/music-news/spotify-ceo-us-launch-is-looking -pretty-good-1211916/.

6.   Glenn Peoples, "Business Matters: Spotify's 2012 Revenues More than Double, Net Losses Increase Indicating Workable Business Model," *Billboard*, July 31, 2013, www .billboard.com/music/music-news/business-matters-spotifys-2012-revenues-more-than -5337666/.

7.   Yukari Iwatani Kane and Russell Adams, "Apple Retreats in Publisher Fight," *Wall Street Journal*, June 9, 2011, https://www.wsj.com/articles/SB10001424052702304392704576375562319751514.

8.   Shayndi Raice, "Facebook Reveals Vulnerability in Mobile," *Wall Street Journal*, May 10, 2012, www.wsj.com/articles/SB10001424052702304203604577394532578783126.

9.   Steven Levy, *Facebook: The Inside Story* (Blue Rider Press, 2021).

CHAPTER 6: AN IRISH PROBLEM, 2013

1.   Citizens for Tax Justice, "Apple Holds Billions of Dollars in Foreign Tax Havens" May 20, 213, ctj.org/apple-holds-billions-of-dollars-in-foreign-tax-havens.

2.   Charles Duhigg and David Kocieniewski, "How Apple Sidesteps Billions in Taxes," *New York Times*, April 28, 2012, https://www.nytimes.com/2012/04/29/business/apples-tax -strategy-aims-at-low-tax-states-and-nations.html.

3.   Darius Dixon, "Coburn: Apple Tax Dodge Is Rotten," *Politico*, May 1, 2012, www.politico .com/story/2012/05/coburn-apple-tax-dodge-is-rotten-075781.

4.   Anna Palmer and Tony Romm, "Apple Defends $100B Offshore Stash," *Politico*, May 16, 2013, www.politico.com/story/2013/05/apple-tim-cook-congress-tax-091501.

5.   Alex Rogers, "Caramel Apple," *Time*, May 22, 2013, swampland.time.com/2013/05/22 /caramel-apple/.

6.   Nelson D. Schwartz and Brian X. Chen, "Disarming Senators, Apple Chief Eases Tax Tensions," *New York Times*, May 21, 2013, www.nytimes.com/2013/05/22/technology/ceo -denies-that-apple-is-avoiding-taxes.html.

7.   Max Colchester, "EU Seeks to Crack Down on Tax Havens," *Wall Street Journal*, Dec. 6, 2012, www.wsj.com/articles/SB10001424127887324640104578162841356823704.

8.   Mehreen Khan, "How EU Forced Ireland and Apple into a €13Bn Tax Defeat," *Times* (London), Sept. 13, 2024, www.thetimes.com/business-money/technology/article/how -brussels-forced-ireland-and-apple-into-a-13bn-tax-defeat-sl7hvzsss.

9.   James Kanter, "E.U. Antitrust Enforcer Will Be Margrethe Vestager," *New York Times*, Sept. 10, 2014, www.nytimes.com/2014/09/11/business/international/eu-antitrust-enforcer -will-be-margrethe-vestager.html.

10. Suzy Hansen, "Meet the Woman Leading Europe's War Against Google, Gazprom, and Apple," *Foreign Policy*, March 18, 2016, foreignpolicy.com/2016/03/18/the-face-of-justice -margrethe-vestager-eu-google-gazprom-antitrust/.

11. Ibid.

12. Press Release, "Commission decides selective tax advantages for Fiat . . ." Oct. 20, 2015, ec.europa.eu/commission/presscorner/detail/en/ip_15_5880.

13. Ryan Heath, "How Vestager Took a Bite out of Apple," *Politico*, Sept. 15, 2016, www .politico.eu/article/how-vestager-took-a-bite-out-of-apple/.

14. Bruce Sewell interview by Dorren Benyamin, "Before You Take the LSAT," (June 9, 2019), https://youtu.be/-wuf3KI76Ds?si=IEwSyXSrsAz7Ngn.

15. Ibid.

16. Heath, "How Vestager Took a Bite out of Apple.".

17. Foo Yun Chee, "U.S. Demands EU Reconsider Tax Probes of Its Companies," Reuters, Feb. 11, 2016.

18. Foo Yun Chee, "EU Antitrust Chief Rejects U.S. Criticism of Apple, Starbucks Tax Cases," Reuters, Feb. 1, 2016.

19. Foo Yun Chee, "EU's Vestager Says Not Penalicing U.S. Firms or U.S. Tax System," Reuters, Feb. 29, 2016.

20. Heath, "How Vestager Took a Bite out of Apple."

21. Foo Yun Chee, "Apple to Appeal EU Tax Ruling This Week, Says It Was a 'Convenient Target,'" Reuters, Dec. 18, 2016.

CHAPTER 7: BATTLE OF THE BANDS, 2014

1.   Tim Ingham, "Global Record Industry Income Drops Below $15Bn for First Time in Decades," MusicBusiness Worldwide, April 14, 2015.

2.   Tim Ingham, "Spotify Revenues Topped $2Bn Last Year as Losses Hit $194M," Music-Business Worldwide, May 23, 2016, www.musicbusinessworldwide.com/spotify-revenues -topped-2bn-last-year-as-losses-hit-194m/.

3.   Yinka Adegoke, "Spotify CEO Daniel Ek on Hitting 10 Million Subs, Apple-Beats Deal, IPO (Q&A)," *Billboard*, May 21, 2014, www.billboard.com/music/music-news/spotify -ceo-daniel-ek-10-million-subs-apple-beats-ipo-6092287/.

4.   Chris Welch, "Spotify Urges iPhone Customers to Stop Paying Through Apple's App Store," *Verge*, July 8, 2015, www.theverge.com/2015/7/8/8913105/spotify-apple-app-store -email.

5.   Joshua Burstein, "Spotify Is Picking a Fight with Apple over the App Store," Bloomberg News, July 8, 2015, www.bloomberg.com/news/articles/2015-07-08/spotify-is-picking-a- fight-with-apple-over-the-app-store?sref=PRBlrg7S.

6.   Brian X. Chen, "Apple Wins Decade-Old Suit over iTunes Updates," *New York Times*, Dec. 16, 2014, www.nytimes.com/2014/12/17/technology/apple-antitrust-suit-ipod-music.html.

7.   Tony Romm, "Spotify Makes Case Against Apple in Congress," *Politico*, July 12, 2015, www.politico.com/story/2015/07/spotify-makes-case-against-apple-in-congress-119976.

8.   David McCabe, "The Trustbuster Who Has Apple and Google in His Sights," *New York Times*, March 22, 2024, www.nytimes.com/2024/03/22/technology/jonathan-kanter -apple-antitrust.html?smid=url-share.

9.   Mia Galupp, "THR's Raising the Bar Honoree: Disney's Disrupter," *Hollywood Reporter*, April 12, 2023, www.hollywoodreporter.com/business/business-news/disneys-disrupter -horacio-gutierrez-1235371228/.

10. Janet I. Tu, "Microsoft's Gutierrez Safeguards Valuable Trove of Patents," *Seattle Times*, Aug. 19, 2012, www.seattletimes.com/business/microsofts-gutierrez-safeguards-valuable -trove-of-patents/.

11. Ina Fried, "Brad Smith Discusses Antitrust Probe's Impact on Microsoft," *Axios*, Sept. 20, 2019, www.axios.com/2019/09/20/microsoft-brad-smith-antitrust-probe-innovation.

12. Peter Kafka, "Spotify Says Apple Won't Approve a New Version of Its App Because It Doesn't Want Competition for Apple Music," Recode, June 30, 2016, www.vox .com/2016/6/30/12067578/spotify-apple-app-store-rejection.

13. Peter Kafka, "Elizabeth Warren Says Apple, Amazon, and Google Are Trying to 'Lock Out' the Competition," Recode, June 29, 2016, www.cnbc.com/2016/06/29/elizabeth -warren-says-apple-amazon-and-google-are-trying-to-lock-out-the-competition.html.

14. Jack Ewing, "Taking a Hard Stance on Microsoft," *BusinessWeek*, March 3, 2006, www .bloomberg.com/news/articles/2006-03-13/taking-a-hard-stance-on-microsoftbusinessweek -business-news-stock-market-and-financial-advice?sref=PRBlrg7S.

CHAPTER 8: SUPER APPS, 2017

1. Kevin Kingsbury, "Apple Is Now More Than Double the Size of Exxon—and Everyone Else," *Wall Street Journal*, Feb. 23, 2015, www.wsj.com/articles/BL-MBB-3357.

2. Lorraine Luk and Daisuke Wakabayashi, "Apple, China Mobile Sign Deal to Offer iPhone," *Wall Street Journal*, Dec. 4, 2013, www.wsj.com/articles/apple-china-mobile-sign-deal-to -offer-iphone-1386207623.

3. Counterpoint, "China: Xiaomi Led the Slow Growing Smartphone Market in 2015," Feb. 18, 2016, www.counterpointresearch.com/insights/china-xiaomi-led-the-slow-growing -smartphone-market-in-2015/.

4. Tim Higgins, "Apple IPhones Sales in China Outsell the U.S. for First Time," Bloomberg News, April 27, 2015, www.bloomberg.com/news/articles/2015-04-27/apple-s-iphones -sales-in-china-outsell-the-u-s-for-first-time?sref=PRBlrg7S.

5. Counterpoint, "China Smartphone Shipments Reached an All-Time High in 2016," Jan. 27, 2017, www.counterpointresearch.com/insights/china-smartphone-shipments-reached -an-all-time-high-in-2016/.

6. Tripp Mickle, "With the iPhone Sputtering, Apple Bets Its Future on TV and News," *Wall Street Journal*, March 25, 2019, www.wsj.com/articles/with-the-iphone-sputtering-apple -bets-its-future-on-tv-and-news-11553437018?st=1koTcZ&reflink=desktopwebshare _permalink.

7. Hannah Frishberg, "Sorry, Android Users: These iPhone Snobs Won't Date You," *New York Post*, Aug. 14, 2019, nypost.com/2019/08/14/sorry-android-users-these-iphone-snobs -wont-date-you/.

8. Alyssa Abkowitz, "The Cashless Society Has Arrived—Only It's in China," *Wall Street Journal*, Jan. 4, 2018, www.wsj.com/articles/chinas-mobile-payment-boom-changes-how -people-shop-borrow-even-panhandle-1515000570.

9. Li Yuan, "A Tip for Apple in China: Your Hunger for Revenue May Cost You," *Wall Street Journal*, May 18, 2017, www.wsj.com/articles/a-tip-for-apple-in-china-your-hunger-for -revenue-may-cost-you-1495100964.

10. Wayne Ma and Juro Osawa, "How Tencent's WeChat Poses Creeping Threat to Apple," The Information, April 14, 2020, www.theinformation.com/articles/how-tencents-wechat -poses-creeping-threat-to-apple?rc=gpeefu.

11. Yoko Kubota and Alyssa Abkowitz, "Apple and Tencent Reach Deal to Let WeChat Users Dole Out Tips," *Wall Street Journal*, Jan. 15, 2018, www.wsj.com/articles/apple-and -tencent-reach-deal-to-let-wechat-users-dole-out-tips-1516018849.

12. "Tencent and Apple Are Negotiating the 'Apple Tax' for Mobile Games," Futubull, Aug. 15, 2024, news.futunn.com/en/post/46569090/tencent-and-apple-are-negotiating-the -apple-tax-for-mobile?level=1&data_ticket=1709667220175636.

13. Wayne Ma, Alyssa Abkowitz, and Julie Steinberg, "Tencent Music, Spotify in Talks to

Swap Stakes Ahead of Planned Listings," *Wall Street Journal*, Dec. 1, 2017, www.wsj
.com/articles/tencent-music-spotify-in-talks-to-swap-stakes-ahead-of-planned-public
-listings-1512122042.

14. Brian Crecente, "Their Future Is Epic: The Evolution of a Gaming Giant," *Polygon*, May
2016, https://www.polygon.com/a/epic-4-0.

15. Wayne Ma and Juro Osawa, "Tensions Flare Behind the Scenes of 'League of Legends,'"
The Information, Aug. 13, 2018, www.theinformation.com/articles/tensions-flare-behind
-the-scenes-of-league-of-legends?rc=gpeefu.

16. Zachery Eanes, "Another Epic Games Exec Joins the Ranks of Triangle Billionaires, Testimony Reveals," *News & Observer*, May 6, 2021, www.newsobserver.com/news/business
/article251169884.html.

17. University of Maryland, "The Epic World of Tim Sweeney," Sept. 9, 2019, https://eng
.umd.edu/news/story/the-epic-world-of-tim-sweeney.

18. WNC staff, "In Land We Trust: Meet 11 Leaders in the Quest to Conserve WNC's
Natural Areas," *WNC Magazine*, May 2014, https://wncmagazine.com/feature/land_we
_trust.

19. Paul Gadi, "Game Design Inspirations: Fortnite's Battle Royale Pivot," Game Developer,
Sept. 19, 2017, www.gamedeveloper.com/design/game-design-inspirations-fortnite-s
-battle-royale-pivot.

20. Kimberly Jenkins, "Creating Fortnite and the Future," BYU alumni spotlight (Spring
2024), science.byu.edu/alumni-spotlight/creating-fortnite-and-the-future.

21. Donald Mustard, X, Dec. 9, 2023, x.com/DonaldMustard/status/1733610758087581706.

CHAPTER 9: FRENEMIES, 2018

1. Shira Ovide, "The Smartphone Revolution Was the Android Revolution," *Bloomberg*,
Aug. 6, 2019, www.bloomberg.com/graphics/2019-android-global-smartphone-growth
/?sref=PRBlrg7S.

2. Gartner, "Gartner Says Annual Smartphone Sales Surpassed Sales of Feature Phones
for the First Time in 2013," Feb. 13, 2014, www.gartner.com/en/newsroom/press
-releases/2014-02-13-gartner-says-annual-smartphone-sales-surpassed-sales-of-feature
-phones-for-the-first-time-in-2013.

3. Jack Nicas and Cade Metz, "Apple Hires Google's A.I. Chief," *New York Times*, April 3,
2018, www.nytimes.com/2018/04/03/business/apple-hires-googles-ai-chief.html.

4. Statista, "Search Engine Market Share of Bing in the United States from January 2018
to March 2024," April 2024, www.statista.com/statistics/1383614/bing-search-engine
-market-share-united-states/.

5. Deepa Seetharaman and Katherine Bindley, "Facebook Controversy: What to Know
About Cambridge Analytica and Your Data," *Wall Street Journal*, March 23, 2018, www
.wsj.com/articles/facebook-scandal-what-to-know-about-cambridge-analytica-and-your
-data-1521806400.

6. Meghann Farnsworth, "Full Transcript: Apple CEO Tim Cook with Recode's Kara Swisher
and MSNBC's Chris Hayes," *Vox*, April 7, 2018, www.vox.com/2018/4/6/17206532
/transcript-interview-apple-tim-cook-msnbc-kara-swisher.

7. Deepa Seetharaman, Emily Glazer, and Tim Higgins, "Facebook Meets Apple in Clash
of the Tech Titans—'We Need to Inflict Pain,'" *Wall Street Journal*, Feb. 13, 2021, www
.wsj.com/articles/facebook-meets-apple-in-clash-of-the-tech-titans-we-need-to-inflict
-pain-11613192406.

8. Jen Kirby, "Mark Zuckerberg on Tim Cook's Criticism of Facebook: It's 'Extremely

Glib and Not Aligned with the Truth,'" *Vox*, April 2, 2018, www.vox.com/technology
/2018/4/2/17183708/mark-zuckerberg-facebook-tim-cook-apple.

9.   Salvador Rodriguez, "The Secret Talks That Could Have Prevented the Apple vs. Facebook War," *Wall Street Journal*, Aug. 12, 2022, www.wsj.com/articles/inside-the-apple-vs
-facebook-privacy-fight-11660317376.

10.   Deepa Seetharaman, Emily Glazer, and Tim Higgins, "Facebook Meets Apple in Clash
of the Tech Titans—'We Need to Inflict Pain,'" *Wall Street Journal*, Feb. 13, 2021, www
.wsj.com/articles/facebook-meets-apple-in-clash-of-the-tech-titanswe-need-to-inflict
-pain-11613192406.

11.   Jem Aswad, "Spotify's Daniel Ek Slams Apple for 'Anticompetitive' Practices in Berlin
Speech," *Variety*, March 14, 2019, variety.com/2019/music/news/spotify-daniel-ek-slams
-apple-anticompetitive-berlin-1203163265/.

CHAPTER 10: A RISING KINGDOM, 2018

1.   Statista, "Registered Users of Fortnite Worldwide From August 2017 to November 2023,"
www.statista.com/statistics/746230/fortnite-players/.

2.   Oscar Gonzalez, "Super Mario Bros. auction breaks record with $660K sale, *CNET*,
April 2, 2021, www.cnet.com/tech/gaming/super-mario-bros-auction-breaks-record-with
-660k-sale/.

3.   Sam Stewart, "Fortnite Accidentally Featured PS4, Xbox One Cross-Platform Play This
Weekend," IGN, Sept. 18, 2017, www.ign.com/articles/2017/09/18/fortnite-accidentally
-featured-ps4-xbox-one-cross-platform-play-this-weekend.

4.   Tim Higgins, "Apple Doesn't Make Videogames. But It's the Hottest Player in Gaming," *Wall Street Journal*, Oct. 2, 2021, https://www.wsj.com/tech/apple-doesnt-make
-videogames-but-its-the-hottest-player-in-gaming-11633147211.

5.   Ibid.

CHAPTER 11: SWEENEY'S REBELLION, 2018

1.   Andrew Martonik, "Epic's First Fortnite Installer Allowed Hackers to Download and
Install Anything on Your Android Phone Silently," Android Central, Aug. 24, 2018,
www.androidcentral.com/epic-games-first-fortnite-installer-allowed-hackers-download
-install-silently.

CHAPTER 14: WILD WEST OF LAW, 2020

1.   Jacob Gershman and Tim Higgins, "In Epic Games v. Apple Trial, an Unorthodox Judge
Presides," *Wall Street Journal*, May 12, 2021, https://www.wsj.com/us-news/law/in-epic
-games-v-apple-trial-an-unorthodox-judge-presides-11620820801.

CHAPTER 16: THE CULTURE WAR, 2021

1.   Abram Brown, "Parler's Founder Explains Why He Built Trump's New Favorite Social Media App," *Forbes*, June 27, 2020, www.forbes.com/sites/abrambrown/2020/06/27
/parlers-founder-explains-why-he-built-trumps-new-favorite-social-media-app/.

2.   S. Matthew Liao, "Do You Have a Moral Duty to Leave Facebook?," *New York Times*, Nov.
24, 2018, www.nytimes.com/2018/11/24/opinion/sunday/facebook-immoral.html.

3.   Mike Isaac and Sheera Frenkel, "Facebook Adds Labels for Some Posts as Advertisers
Pull Back," *New York Times*, June 26, 2020, www.nytimes.com/2020/06/26/technology
/facebook-labels-advertisers.html.

4.   Katie Glueck, Michael S. Schmidt, and Mike Isaac, "Allegation on Biden Prompts Push-back from Social Media Companies," *New York Times*, Oct. 14, 2020, www.nytimes .com/2020/10/14/us/politics/hunter-biden-ukraine-facebook-twitter.html.

5.   Kate Conger and Mike Isaac, "Twitter and Facebook Labeled Posts from Trump, but Only One Moved to Limit Their Spread," *New York Times*, Nov. 3, 2020, www.nytimes.com /live/2020/11/03/us/trump-biden-election/twitter-and-facebook-labeled-posts-from -trump-but-only-one-moved-to-limit-their-spread?smid=url-share.

6.   Sleeping Giants tweet, Jan. 7, 2021, x.com/slpng_giants/status/1347190280492089344.

7.   Ryan Mac and John Paczkowski, "Apple Has Threatened to Ban Parler from the App Store," *BuzzFeed*, Jan. 8, 2021, www.buzzfeednews.com/article/ryanmac/apple-threatens -ban-parler.

8.   Jack Nicas and Davey Alba, "How Parler, a Chosen App of Trump Fans, Became a Test of Free Speech," *New York Times*, Jan. 10, 2021, www.nytimes.com/2021/01/10/technology /parler-app-trump-free-speech.html.

9.   www.foxnews.com/politics/nunes-racketeering-investigation-amazon-apple-google -parler-ban.

10. ndlegis.gov/assembly/67-2021/testimony/SIBL-2333-20210209-6054-A-NEUENSCHWANDER_ERIK.pdf.

11. Jack Nicas, "Big Tech's Unlikely Next Battleground: North Dakota," *New York Times*, Feb. 14, 2021, www.nytimes.com/2021/02/14/technology/north-dakota-tech-apps.html.

12. Kif Leswing, "Apple Wins Victory as North Dakota Votes Down Bill That Would Regulate App Stores," CNBC.com, Feb. 16, 2021, www.cnbc.com/2021/02/16/apple-wins-victory -as-north-dakota-votes-down-bill-that-would-regulate-app-stores.html.

13. Tech Transparency Project, "Inside Apple's Push to Kill State App Store Bills," Sept. 7, 2021, www.techtransparencyproject.org/articles/inside-apples-push-kill-state-app-store-bills.

14. Ryan Tracy and Tim Higgins, "Apple Finds Itself Under Scrutiny in Washington's Big Tech Clampdown," *Wall Street Journal*, Feb. 20, 2022, www.wsj.com/articles/apple-finds -itself-under-scrutiny-in-washingtons-big-tech-clampdown-11645353001.

CHAPTER 17: INFINITE LOOP(HOLES), 2021

1.   Jiyoung Sohn, "Google, Apple Hit by First Law Threatening Dominance over App-Store Payments," *Wall Street Journal*, Aug. 31, 2021, www.wsj.com/articles/google-apple-hit -in-south-korea-by-worlds-first-law-ending-their-dominance-over-app-store-payments -11630403335?mod=e2tw.

2.   Jiyoung Sohn, "Google Allows Alternate In-App Payment Options in South Korea, Though Familiar Fees Remain," *Wall Street Journal*, Nov. 4, 2021, www.wsj.com/articles /google-allows-alternate-in-app-payment-options-in-south-korea-though-familiar-fees -remain-11636021788.

3.   Stephanie Bodoni, Mark Gurman, and Aoife White, "Apple Cuts Fees for Dating Apps in Dutch Antitrust Response," Bloomberg News, Feb. 2, 2022.

4.   Aoife White, "Apple Racks Up More Fines in Dutch Fight over Dating-App Payments," Bloomberg News, Feb. 14, 2022, www.bloomberg.com/news/articles/2022-02-14/apple -racks-up-more-fines-in-dutch-spat-over-dating-app-payments?sref=PRBlrg7S.

5.   Margrethe Vestager, "Shared Objectives for Framing the Tech Economy," Berkeley, California, Feb. 22, 2022, lawcat.berkeley.edu/record/1266833?v=pdf.

6.   Valentina Pop and Rochelle Toplensky, "EU Commissioner Who Targeted Tech Giants Gets Second Term," *Wall Street Journal*, Sept. 10, 2019, www.wsj.com/articles/eu -commissioner-who-targeted-u-s-technology-giants-gets-second-term-11568119303.

7. Marie-Louise Truelsen and Susanne Baden Jensen, "Margrethe Vestager: Min man dog datter er flyttet tilbage til Danmark," *Alt*, April 4, 2018, alt.dk/artikler/margrethe-vestager -min-mand-og-datter-er-flyttet-tilbage-til-danmark/2787436.

8. Valentina Pop and Sam Schechner, "Apple Wins Major Tax Battle Against EU," *Wall Street Journal*, July 15, 2020.

9. Pietro Lombardi, "Big Tech Boosts Lobbying Spending in Brussels," *Politico*, March 22, 2022, www.politico.eu/article/big-tech-boosts-lobbying-spending-in-brussels/.

10. Corporate Europe Observatory, "The Lobby Network: Big Tech's Web of Influence in the EU," Aug. 31, 2021, https://corporateeurope.org/en/2021/08/lobby-network-big-techs -web-influence-eu.

CHAPTER 18: RETURN OF SWEENEY, 2023

1. Nico Grant, "Google Must Open Android to Other App Stores and Billing Options, Judge Rules," *New York Times*, Oct. 7, 2024, www.nytimes.com/2024/10/07/technology/google -app-store-epic-games-anticompetitive.html.

2. Kif Leswing, "Epic Games CEO Tim Sweeney on Why the Company Did Better Against Google Than Apple in Court," CNBC.com, Dec. 12, 2023, www.cnbc.com/2023/12/12 /tim-sweeney-why-epic-did-better-against-google-than-apple-in-court.html.

CHAPTER 19: THE IDES OF MARCH, 2024

1. Jay Peters, "Apple's App Store Policies Now Let U.S. Developers Link to Outside Payments," *Verge*, Jan. 16, 2024, www.theverge.com/2024/1/16/24040881/apple-outside -payments-app-store-policies-iphone-ipad.

2. Mike Scarcella, "Apple Asks US Judge to Toss App Store Injunction," Reuters, Sept. 30, 2024, www.reuters.com/legal/transactional/apple-asks-us-judge-toss-app-store -injunction-2024-09-30/.

3. Adam Satariano and Tripp Mickle, "Apple Is First Company Charged Under New E.U. Competition Law," *New York Times*, June 24, 2024, www.nytimes.com/2024/06/24 /technology/apple-european-union-competition-law.html.

4. Kim Mackrael, "Apple Fined $2 Billion in One of Europe's Largest Antitrust Actions," *Wall Street Journal*, March 4, 2024, www.wsj.com/tech/apple-hit-with-near-2-billion-fine -in-europe-over-music-streaming-apps-74062ff7.

5. Aude Van Den Hove, "European Commission Scores Stunning Court Win in €13B Apple Tax Row," *Politico*, Sept. 10, 2024, www.politico.eu/article/commission-scores-surprise -win-in-apple-tax-row/.

6. Jan Wolfe, "The Google Antitrust Verdict Looms. Here's What to Look For," *Wall Street Journal*, May 5, 2024, www.wsj.com/tech/the-google-antitrust-verdict-looms-heres-what -to-look-for-804065c1.

7. Josh Sisco, "The Top Biden Lawyer with His Sights on Apple and Google," *Politico*, Jan. 18, 2023, www.politico.com/news/2023/01/17/an-antitrust-revival-dojs-kanter-takes-big -swings-and-misses-to-fight-monopolies-00077304.

INDEX

ABOUT THE AUTHOR

Tim Higgins is the author of *Power Play* (about the rise of Tesla). A frequent CNBC contributor, he writers about Silicon Valley for the *Wall Street Journal* and lives in San Francisco.